AF616233

Thick Film Circuits

Thick Film Circuits

G. V. PLANER, PH.D., M.SC., F.R.I.C.
Director, G. V. Planer Ltd.

L. S. PHILLIPS, B.SC., M.A.
Senior Executive Scientist, G. V. Planer Ltd.

LONDON BUTTERWORTHS

THE BUTTERWORTH GROUP

ENGLAND

Butterworth & Co. (Publishers) Ltd.
London: 88 Kingsway, WC2B 6AB

AUSTRALIA

Butterworths Pty Ltd.
Sydney: 586 Pacific Highway, Chatswood, NSW 2067
Melbourne: 343 Little Collins Street, 3000
Brisbane: 240 Queen Street, 4000

CANADA

Butterworth & Co. (Canada) Ltd.
Toronto: 14 Curity Avenue, 374

NEW ZEALAND

Butterworths of New Zealand Ltd.
Wellington: 26–28 Waring Taylor Street, 1

SOUTH AFRICA

Butterworth & Co. (South Africa) (Pty.) Ltd.
Durban: 152–154 Gale Street

First published 1972

ISBN 0 408 70395 4

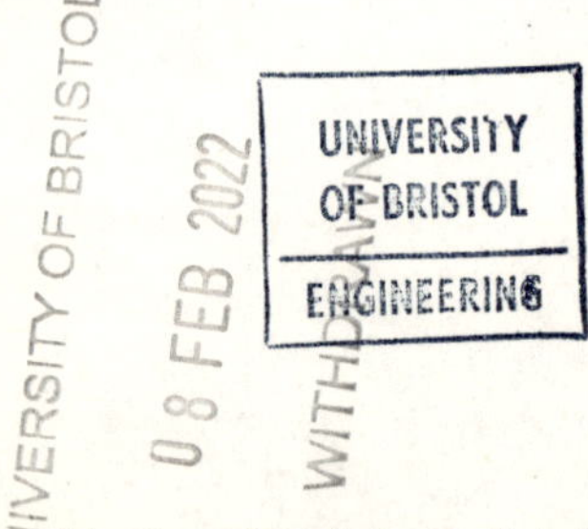

Printed and bound in Great Britain by R. J. Acford Ltd.,
Industrial Estate, Chichester, Sussex

Preface

Hybrid microelectronic assemblies offer one of the most interesting and important challenges to the electronics based industries of today. The number of applications is rapidly increasing and it seems that most, if not all, areas eventually will benefit from the impact of this composite technology. Thick film techniques have now emerged as a most important factor in microelectronic developments, and by virtue of their versatility and simplicity, seem destined to provide the basis of hybrid microcircuit manufacturing procedures for some time to come.

This book has been written primarily to provide an understanding of the materials and techniques involved for those using, or contemplating the use of, thick film circuits, but who have had no experience in microelectronics. In addition, however, it should provide a useful reference book for those already involved in this area, or who have a more general interest in electronic packaging developments.

Our readers' likely wide range of technical backgrounds has presented difficulties in determining the style of presentation in some areas. We have worked on the assumption, supported by our own widespread contacts in the industry, that, while not always interested in the theoretical aspects, most of those involved do regard it as important to understand the nature of the materials with which they work. Similarly there is often considerable interest in how processing conditions may be manipulated to achieve specific results, whether in terms of physical configuration or performance characteristics.

A study of the contents page will show that we have chosen a direct and practical approach which we feel will be most appreciated by those for whom this book is intended. After an introduction placing thick films in context and an explanation of the range of current and potential applications, the manufacturing procedures are covered step by step from the bare substrate through to the complete packaged and tested circuit. Having en-route discovered the limitations of the technology, we then provide a chapter laying out the design concepts involved in the translation of typical circuits from paper to thick film. Finally, we have a section on 'Trends and Future Developments' in which we try to guess far enough ahead to outstrip the present pace of progress. We hope that we shall have succeeded!

The thoughts and work of many persons have been gathered to write this book and it is of course impossible for us to acknowledge all those to whom our thanks are due. However, we should like to mention especially Mr. P. J. Evison, Dr. V. J. Hammond and Mr. D. F. A. MacLachlan, of G. V. Planer Ltd., whose co-operation and assistance have contributed to the completion of this book. To those acknowledged above, to those referenced in the text and bibliography and to those inadvertently unmentioned, we are extremely grateful.

G. V. Planer

L. S. Phillips

Contents

1

Introduction

The main stream of microelectronic technological development is composed of a number of closely interacting currents, the most important of which at the present time involve the thick film, thin film and monolithic semiconductor systems. Each of these has, in the past, been projected as the universally applicable circuit technology, and, whilst each has certain specially attractive features, it now appears that a highly flexible hybrid technology combining these three approaches will emerge as the electronic/electrical engineering work-horse for some years to come.

1.1 DEVELOPMENT OF CIRCUIT FABRICATION TECHNOLOGIES

The military and defence requirements arising from the Second World War and the ensuing 'cold' war provided the initiating drive which set in motion rapid development of the miniature, high reliability electronic and electrical systems which we now take for granted. The urgent requirement for ever greater circuit complexity, stability, lightweight and fully functioning capabilities in an aero-spatial environment provided the impetus and made available the large sums of money necessary to launch the dramatic acceleration of materials and device fabrication technology that we have seen.

Key factors in the early stages of these developments were the emergence of printed circuit board technology, together with the availability of miniature electronic valves and passive components.

Next the transistor made its appearance, immediately affording substantial savings in volume, reducing thermal dissipation problems and, by virtue of the low voltage operation possible, allowing further reduction in size of components such as capacitors.

Thin film microcircuits began to appear at this time; these circuits deposited by vacuum deposition onto glass and ceramic substrates were shown to be capable of providing resistor and capacitor film devices with considerable precision and stability under laboratory conditions. It was therefore projected that 'integrated circuits' of this type incorporating thin film resistors, capacitors, inductors and eventually active transistor-type devices, would provide an answer to most circuit fabrication problems.

Subsequently, the thick film screen printing system was developed, claiming, where extremes of accuracy and stability were not required, to be considerably simpler in processing and of low cost in production.

More recently we have seen the general availability of multi-device and multi-function semiconductor integrated circuits providing very high component density and, where large numbers are required or where user oriented programming is feasible, very low cost per function.

In parallel with these other developments the printed circuit board has made steady, if perhaps unspectacular, progress. At the present time, due to the considerable improvements in board quality and innovations such as plated through holes and multilayering, it still holds a major place as a general purpose low cost circuit carrier system. Utilising the new ranges of miniature components in conjunction with, for example, dual in-line packaged semi-conductors, it is possible to provide a very flexible approach to circuit interconnection and packaging problems.

1.2 COMPETING CIRCUIT TECHNOLOGIES

Each of the systems mentioned has its strengths and weaknesses. Before discussing in greater detail the reason for the considerable interest in thick film circuitry at the present time, we shall examine a little more closely the individual circuit technologies in order to illustrate the central position which the thick film approach currently holds.

1.2.1 PRINTED CIRCUIT BOARD

Printed circuit board consists basically of thin sheets of copper conductor pressed on to a resin-bonded substrate material, the copper being etched chemically to leave a network of conductive areas providing interconnections and contact lands for the attachment of discrete components. These components are normally attached by passing their leads or supporting pins through holes in the board for soldering to conductive areas on the reverse side of the board.

As components have become smaller and efforts have been made to crowd them closer together, and particularly with the advent of multiple lead devices such as the now widely employed 14-pin dual in-line packages, the problems of producing sufficiently narrow and suitably routed conductor patterns have grown considerably more severe. This has to a large extent been overcome by the use of double-sided and multilayered boards in which a number of conductor planes are stacked with intervening insulating layers upon one substrate board. Interconnection crossovers are made by electroplating through-holes to join the various conductive planes at the appropriate positions. These techniques, however technically satisfactory, do mar the basic attraction of simplicity in printed circuit boards and considerably increase costs, again previously a major attraction of the printed circuit board approach.

Another limitation which has emerged with the progressive tendency towards higher circuit density is the difficulty of dissipating heat generated by active components, resistors etc. in operation. When anchored to a metal chassis or attached to relatively large expanses of copper conductor foil, adequate heat sinking capacity is normally available. But as conductors become thinner to provide high interconnection density, and due to the relatively poor thermal conductivity of the resin-bonded substrate, destructively high temperatures may be reached in operation. The use of metal-cored substrate boards has been explored in special applications, but apparently so far without sufficiently promising results to warrant, at least for general purposes, the additional costs involved.

1.2.2 THIN FILM CIRCUITS

Thin film circuits are prepared by boiling material under reduced pressure and allowing the vapour so produced to deposit onto a substrate, usually of glass or ceramic. Two main approaches are employed: the use of masks with apertures allowing deposition of the material only over the exposed areas to provide the desired components pattern; or selective etching, in which process the whole of the substrate surface is coated and the required pattern is subsequently produced by normal photoresist-etch techniques to remove the unwanted areas of the coating. Materials employed here for film formation may be gold or aluminium for conductors, tantalum and nickel–chromium alloys for resistor patterns and silicon oxides, glass, tantalum oxides, etc. for dielectric and insulating films.

Advantages of the thin film system are that precisely defined patterns and components can be produced, for instance, resistors deposited on the same substrate will have closely similar characteristics.

It is possible to produce high density circuits which are intimately bonded to the substrate surface and hence have good heat transfer characteristics, as compared with printed circuit board assemblies in which only the component leads are normally attached to a body capable of conducting heat away from the components. On the other hand, the film components being of small thickness (typically only a few thousand angstroms), it is not normally possible to fabricate and operate high wattage dissipation devices or systems. However, in the context of today's trend towards lower power requirements for transistors and integrated circuits, this is not always a limitation.

Disadvantages of the thin film approach are mainly in the fabrication; capital costs are relatively high, and skilled personnel are often required to ensure reasonably high yields of circuits produced by this method.

1.2.3 SEMICONDUCTOR INTEGRATED CIRCUITS

Semiconductor integrated circuit techniques have made most impressive progress over the last few years. Hardly had the transistor become established as a normal professional and industrial component before the first generation of multi-device functional semiconductor units made their appearance.

In these integrated circuit chips, consisting basically of high purity silicon, the process of selective adjustment of the electrical characteristics of the crystal by doping to produce pn junctions as in normal diodes and transistors is extended to provide discrete areas having resistive and capacitive characteristics.

Thus it is possible in principle to produce almost any circuit configuration in monolithic form by careful design, provided that the necessity for employing certain 'difficult' components such as inductors and large value capacitors, for example, can be avoided. The extremely small size of each individual component area allows very large numbers of 'components' to be packed on small chips of semiconductor material measuring typically 0.04–0.10 in^2.

Where there is sufficient volume of demand to warrant it, long runs of standard items such as logic gate functions, shift registers, memory cells etc. have made it possible to bring the price per function (which may employ several integrated transistors, diodes, resistors, capacitors etc.) down to a level similar to that previously obtained for a single discrete transistor. This is, of course, very attractive to the designer, providing him with a sophistication of circuitry undreamt of only a few years ago in a very small space and a relatively low cost. However, two important limitations to the development and applications of integrated circuits are: the problems associated with interconnecting microdevices of this type, and the high cost and relatively long design cycle involved in circuit origination and modification.

At the time of writing it has not yet proved possible, except in some limited instances, to provide a chip carrying a complete circuit system so that it merely needs to be connected to power supply leads to be fully functional. Thus, most integrated circuits must be interconnected with various other discrete components and systems. Here the problem of thermal dissipation again occurs and it has proved necessary in some instances to mount the chips on a metal tab to provide sufficient heat sinking. This is a problem which will become progressively aggravated by the present trend to pack more and more functions and devices onto a chip at ever greater densities.

Probably the most commonly employed method of interconnecting monolithic chips is by back-bonding to a carrier plate, followed by the attachment of jumper leads from the contact areas on the chip to those on the carrier plate. This process is time consuming and hence expensive, and the large number of wire bonds involved inevitably degrades the maximum attainable level of reliability for a particular system.

The practice of 'flip-chipping' of devices, involving attachment of the contact areas directly to corresponding areas on the carrier, should considerably improve the reliability from the bonding point of view. However, it will markedly worsen the thermal problems, as only a relatively small proportion of the chip area will be in contact with a heat sink. An intermediate solution now coming into general practice is the use of beam lead devices in which gold beams intimately attached to the semiconductor chip are bonded to relatively large contact pads on a mother board.

Another disadvantage of semiconductor integrated circuits lies in the fact that, despite considerable efforts, it has proved impossible to produce other than very limited ranges of resistance and capacitance values on the chip.

Probably the most significant drawback for the majority of circuit users, who require relatively small numbers of many different types of circuits, as for example, manufacturers of instrumentation and special control systems, incur the significant drawback in the high cost of installing a production facility where 'in-house' operation is desirable. Additionally they have to bear the costs of originating the complex designs, diffusion masks and processing schedules to implement a new circuit.

1.2.4 THICK FILM CIRCUITRY

Thick film technology offers a flexible low cost answer to the more specialised, small volume requirements of users. In this approach pastes are screen printed by conventional methods onto a ceramic substrate to define conductor and resistor patterns; firing under the appropriate conditions of time and temperature produces rugged components bonded to the substrate and protected in a glassy matrix. The capital outlay and running costs are relatively low, highly qualified or skilled operators are not essential and changes in circuit design are readily and inexpensively implemented in production.

The resistive, thermally dissipating components are bonded directly to a substrate (usually alumina) and so are capable of much higher wattage dissipation without degrading effects. A range of low value capacitors may be printed if required, but other larger value capacitors and a very wide range of discrete devices, both active and passive, may be readily attached to conductive areas on the circuit. This may be done by soldering, without damaging the circuit or other components. Since the conductor films are appreciably thicker than for the thin film case discussed previously (of the order of 0.001 in), it is possible for a given maximum permissible series resistance, to go to appreciably narrower conductor lines and so facilitate the attainment of higher circuit densities.

As will be seen, the thick film approach overcomes most of the disadvantages inherent in other systems, provided that extremely small size, or very high accuracy and stability are not vital requirements.

If this thick film technology is so advantageous and so inexpensive, why has it only recently come into general use? The answer is, of course, that the basic materials and ideas of thick film technology have indeed been with us for some time, but only over the last few years has the requirement for this technology reached a sufficient intensity and volume to make it a practical, commercial proposition. The movement towards thick film hybrids has been accelerated by the gradual realisation by users that semiconductor integrated circuits will not be so universally applicable as was first indicated.

1.3 THICK FILM TECHNOLOGY

Let us examine in outline then the actual materials, equipments and processes involved in this most flexible approach to micro-circuitry. The substrate employed is almost invariably a 96% alumina with a surface finish of approximately 25 microinches. Onto these sheets, which are available in a range of standard sizes e.g. 1 in × 1 in, 1.2 in × 0.8 in, 4 in × 4 in and larger, with thickness usually of the order of 0.02–0.04 in, a pattern of conductor paste containing precious metals, is printed through a metal or nylon screen.

The patterns are dried and then fired to a high temperature in a carefully regulated furnace to develop the appropriate conductivity of the film and to bond it integrally with the surface of the substrate. Subsequently, employing appropriately selected 'resistor' pastes, a pattern of resistive areas is printed onto the plate in such a configuration as to be contacted by areas of the precious metal conductor pattern previously laid down at the required locations. These patterns are then dried and fired as before to provide the appropriate resistor characteristics and to bond them intimately to the substrate surface.

Where a tight tolerance on resistor value is required, resistors may be trimmed at this stage, usually by abrading away part of the resistive tracks, to raise the values into the desired range. Where desired, capacitors can be printed by overlaying part of the conductor pattern with a printed film of a dielectric, followed by a further superimposed counter-electrode material, again connected appropriately to the conductor pattern. The dielectric may be glass for small value capacitors, or special glaze compositions containing, for example, barium titanate, where higher values are required.

Discrete components such as active devices or large value capacitors may be readily added to the circuit, e.g. by soldering to contact areas provided at appropriate sites in the contact pattern. The circuit may now be given a coating of a resin for general protection. Improved protection may be obtained through encapsulating, by transfer moulding, or in a hermetically sealed metal or ceramic enclosure to give a completely protected circuit assembly.

The above is only a very brief outline of the procedures which are to be described in much greater detail in later chapters of this book.

1.3.1 ADVANTAGES OF THICK FILM TECHNIQUE

The reasons for the success and rapid expansion of thick film utilisation in industry can be summarized as follows:

1. Growth of demand for relatively small numbers of similar but not identical types of circuit.
2. The capital and running (i.e. labour and materials) costs are sufficiently low to make the setting up of an 'in-house' production facility attractive even to small users.
3. The user and manufacturer can choose from a wide range of circuit complexities, that is from simple conductor patterns to highly complex multilayer mother substrates carrying integrated circuits plus other discrete active and passive devices, whilst using the same basic materials and equipment.
4. Circuit designs may be rapidly and economically implemented in production even by the small vloume producer.
5. For a given facility, production volume can be readily varied over a wide range whilst still maintaining economic operation.
6. The manufacturing process is relatively simple, involving, largely, well established procedures.
7. Where cost is not of prime importance, it is possible to adjust the thick film process to give precision comparable with that of thin film techniques.
8. Since an 'in-house' facility is economically feasible for smaller companies, it is possible in these cases to avoid problems of quality control, delivery and security, which may arise from dependence on subcontractors for supply of circuits.

The thick film circuit is now entering a phase in which it can be expected to progressively displace the printed circuit board system in many applications. It can achieve this since it need be no more expensive if produced in reasonable volume. It is more electrically and mechanically stable in operation, particularly at high temperatures, and since a large proportion of thick film components will be integral with the substrate, higher power dissipations are permissible. Moreover, the thick film circuit is more compact and rugged than its printed circuit board equivalent, it provides higher reliability and is simpler for 'in-house' production by small manufacturers for short runs.

1.3.2 THICK FILM APPLICATIONS

The uses and users of thick film microcircuit technology are already very widespread and increasing in number daily. Probably the most demanding areas of performance are in the aerospace and military applications for which ruggedness, stability and reliability are at a premium. Here, the requirements for minimal volume and weight are crucial, with cost being relatively unimportant.

Another intensive area of use is in computers, where the very large numbers of repetitive circuits required, enable the cost per unit to be held at a low level, whilst still retaining the advantages of relatively uncritical processing to produce units of high stability and reliability.

A growing area of use also is in the manufacture of precision instruments and control systems. Here we have a wider range of performance characteristics versus cost criteria, but in general the extremes of ruggedness and stability are not required and, whilst the low cost is important, it can often be traded for certain improvements in characteristics. At the other end of the scale, we have applications such as the automobile industry and domestic electrical equipments, in which cost will be the most crucial consideration.

The thick film hybrid approach to the manufacture of electronic and electrical equipments offers a very wide range of interesting opportunities and we shall in the next chapters discuss more specifically the extensive range of applications for this technology. We shall also detail some of the many alternative choices in materials and methods available for the realisation of circuits and systems by means of this very flexible approach to circuit fabrication.

2

Applications of Thick Film Circuits

As is common at the introduction of most new technologies, the pioneers of thick film usage saw in this technique a universal answer to the whole range of circuit fabrication problems. Indeed, viewed against the historical background of discrete component technology, it is clear that thick films appear very attractive and capable of providing an approach leading to considerable improvements in the performance and economics of microcircuit systems. Thus the initial phase of thick film technology has seen the fabrication of a very wide range of circuit types, often revealing remarkable ingenuity and precision in manufacture.

However, whilst the practicable spectrum of thick film applications is today considerably wider than when the technology was first introduced, the range of actual and worthwhile usage is currently conditioned to a large extent by the relative merits of the competing semiconductor integrated circuit, and to a lesser extent of the printed circuit board and thin film systems. As an example of this we have the companies who in the early 1960s pioneered the large scale use of thick film circuits in digital logic for computers.

The first generation of circuits comprised largely screen printed components with added transistors in flip-chip form. Recent years, however, have seen the gradual transition through medium scale integration (MSI) to large scale integration (LSI) chips. Now, in this particular application, the thick film circuit has been largely reduced to a mother-plate/interconnect role, as the improved performance and falling cost of semiconductor integrated circuits has tilted the balance in this direction.

There are many specific areas in which the thick film hybrid approach has distinct advantages to offer over the alternative circuit systems. However, there are other fields, particularly those in which semiconductor integrated circuits excel, in which thick films must complement and compete. Thus one method of defining some of those areas in which thick film applications have most to offer is to study those performance areas in which semiconductor integrated circuits do not adequately cover present circuit requirements. We shall discuss these applications areas at component level in more detail.

2.1 COMPONENTS—ACTIVE DEVICES

In the production of semiconductor integrated devices it is usually difficult to hold close performance tolerances, due to the sensitivity of performance characteristics to small changes in materials and/or processing conditions. Hence the initial concentration of integrated circuit manufacture on digital circuits, where 'on/off' rather than graded response characteristics are required. In thick film hybrid circuits where active devices are added as discrete elements, such problems are completely avoided, since chips or micropackaged devices which have been preselected and tested to any required tolerance, can be employed.

It is not usually practicable to provide in a single integrated circuit a wide variety of active devices, such as pnp and npn transistors, diodes, zeners, field effect devices, resistors, capacitors etc. The large number of process steps involved results in low overall yield and hence higher cost. On the other hand, thick film hybrid circuits can carry as wide a range of devices as are available in suitable form for mounting; moreover, in each case the devices can be individually pretested before mounting (and removed after mounting if found to be faulty) with little effort.

There are quite severe limitations on power dissipation in integrated circuit components, due to the high density of packing and relative inefficiency of practicable cooling methods. Thick film microcircuits in which devices are mounted on a high thermal conductivity alumina or insulated metal substrate are much better placed from this point of view. Thick film passive components are also markedly less sensitive to thermal damage.

Integrated semiconductor devices are, generally speaking, limited in voltage handling capacity by the method of insulation employed between various 'component' areas. In the case of high density bi-polar devices, this is currently limited to a few volts by the breakdown voltage of the back-biased pn junction widely employed as isolation at the present time. The advent of Metal Oxide Silicon (MOS) devices and dielectric isolation techniques may considerably improve the situation here. Discrete devices, either as chips or micropackages as incorporated in thick film hybrid circuits, may be mounted on highly insulating alumina substrates, and may currently have voltage ratings as high as several hundred volts for transistors, or thousands of volts for SCR and similar devices.

2.2 COMPONENTS—PASSIVE DEVICES

Integrated semiconductor resistors are normally limited in value to less than 10 kΩ by the difficulty of obtaining sufficiently high specific resistivity by doping of silicon under circuit manufacturing conditions. This is to some extent being overcome by the deposition of thin film resistors onto the silicon substrate. In contrast, the thick film techniques can be used without difficulty to produce resistors ranging in value from a few ohms to many megohms.

In integrated circuits, capacitors are normally limited to values of less than 100 pF and have low Q values. However, the situation here is improving with the availability of MOS capacitors. In thick film hybrid circuits capacitors with values of up to a few thousand pF may be deposited by screen printing. But at the present time it is more economical, where higher values are required, to attach capacitor chips of various types, for example ceramic, mica or tantalum, which provide values of up to several microfarads with a choice of characteristics.

In thick film circuits small value inductors can be screen printed, or miniature inductors can be attached directly to the circuit. The fabrication of inductors in semiconductor integrated form has so far not seemed practicable. Various suggestions for alternative means of simulating inductance have been put forward, such as the use of gyrator circuits, inductively phased diode arrays, or the use of resonant beams in tuned circuits.

Precision of semiconductor resistors and capacitators in integrated form is generally poor, and, furthermore, there is no fully satisfactory method available for adjusting the values once produced. On the other hand thick film components of this type may be adjusted to any accuracy required, or alternatively precalibrated components can be mounted directly upon the substrate.

Integrated resistors and capacitors normally have relatively high temperature coefficients of performance, whereas for thick film hybrids, materials are now available for screen printing of components with low temperature coefficients. Alternatively, selected discrete components may be added to provide the whole range of positive, negative and zero coefficients as required.

Semiconductor integrated resistors and capacitors are quite strictly limited in respect to voltage, power and thermal dissipations, particularly where back-biased junction isolation techniques are employed. Devices in hybrid circuits, on the other hand, may be selected to provide optimum performance in the particular respect required by employing the appropriate technology in each instance.

Thick film components show a greater resistance to damage by intense electromagnetic and particle irradiation than do semiconductor devices.

Thick film hybrid circuits are not limited to the use of conventional resistor, capacitor and silicon semiconductor devices but, in contrast to the integrated circuit systems, it is possible to incorporate a much wider range of 'non electronic' components. These may use magnetic, optical and ferroelectric effects to achieve their functions. In addition it is, of course, possible to employ the more specialised semiconductor properties of germanium, gallium arsenide, indium antimonide and cadmium sulphide in optimised devices.

2.3 SYSTEMS CONSIDERATIONS

Apart from the more direct technical advantages at component level set out above, the following aspects must often be considered when weighing up the relative merits of the microcircuit alternatives available for a particular application.

Thick film hybrid technology is available as an 'in-house' facility to small concerns due to the relatively low capital investment required, and to the fact that highly skilled operators are not essential. The process is economical even for short runs, has a short design-to-production cycle time and lends itself to ready converstion of the design from printed circuit board layouts. An additional advantage lies in the commercial/military security afforded by an in-house facility, together with the ability to retain full control over design, quality control and delivery schedules, as opposed to dependence upon outside suppliers.

It is possible to provide multipackage precision components not otherwise available in high density form. Particular applications for example include precision diode and resistor arrays for switching and decoding.

Thick films on ceramic and insulated metal substrates are replacing printed circuit board systems in applications where the higher temperature/power operation allowed by the former makes the effort of changing technologies worthwhile. A further advantage of thick film circuits in this application is provided by the possibility of adding non packaged chips (not practicable in the case of the printed circuit based systems) to give much higher circuit density.

The thick film hybrid form is particularly suited for the application of cooling techniques. The components are normally mounted on a high thermal conductivity material (e.g. alumina) which may be soldered or brazed to metal conductors to further enhance cooling effects. Evaporative cooling techniques employing, for example, fluorinated hydrocarbons, 'Heat Pipe' and thermoelectric cooling methods are also readily applicable.

The availability of an 'in-house' thick film circuit facility and technology provides the manufacturer with the opportunity of producing, in addition to the resistor networks referred to above, variable resistors and potentiometers employing cermet resistive elements. There are a number of manufacturers producing such components at the present time, and advantages claimed are durability of track, low cost, and essentially infinite resolution.[27,28,29,30] Some difficulties have been encountered with regard to noise (largely due to the nature of the resistive track where highly conductive metal particles are embedded in an insulating glassy matrix), and this has to some extent limited usage to trimmer type applications.

2.4 TYPICAL CIRCUIT APPLICATIONS

Turning now to more specific circuit applications, let us first list some published examples, before proceeding to a more detailed analysis of the reasons for selecting the thick film fabrication approach in typical cases.

COMPUTERS, INSTRUMENTATION AND CONTROL

Digital systems. Gates, switches, adders, shift registers, memory core drivers, sense amplifiers, displays, lamp drivers, relay drivers, clocks, clock drivers, voltage/level shifters, digital to analogue converters, analogue to digital converters, paper tape/card readers, stabilised power supplies.[13,15,16,19,24,26,34,48–51,54,63,65]

Analogue Systems. High performance operational amplifiers, gates, switches, multipliers, dividers, analogue to digital converters, digital to analogue converters, active filters, servo-amplifiers, displays, photosensitive arrays, stabilised power supplies.[13,19,26,32,48–51,73,74,76]

COMMUNICATIONS

TV and radio. UHF/VHF tuners, amplifiers and oscillators, RF tuners, amplifiers and oscillators, modulators, demodulators, decoders, attenuators, active filters, delay modules, colour signal processing systems, stabilised power supplies.[4,16,18,19,32,35,46,47,53,56,58,59,73,74,76]

Audio. Amplifiers, attenuators, public address systems, stereo decoders, 'entertainment' systems for automobiles, sonar.[11,12,20,33,56,59,62,64,66]

Telephony. Amplifiers, attenuators.[62]

Microwaves. Microstrip conductors, couplers, multiplexers, modulators, mixers, delay modules, phase shifters, attenuators, detectors.[6,46,47,54,61,65,69–72,76]

POWER SUPPLIES

Alternators, stabilised systems, filters, choppers, voltage regulators, fuses, level shifters.[8,10,31,52,60,75]

MULTILAYERS AND INTERFACES

Mother board/Interconnect system in high density semiconductor integrated circuit assemblies, digital/analogue interfaces, voltage level shifters.[34,36–43,50,76]

DISCRETE COMPONENTS

Elements for fixed and variable resistors and potentiometers.[27–30,57,68]

In the field of digital logic circuits, and especially for computers, semiconductor integrated devices reign almost supreme and, for general purpose systems, seem unlikely to be displaced within the foreseeable future. There are some areas, however, even in this highly competitive field, where thick film techniques can capitalise on their particular advantages.

With the tendency for more and more devices per semiconductor chip and ever closer packing of chips, the requirement for improved cooling rapidly increases. The thick film conductor pattern on a high thermal conductivity alumina substrate begins to have considerable advantages, as a high density carrier system, over printed circuit board in which the conductor pattern is mounted on a relatively poor thermal conductivity material, such as glass filled epoxy or silicone resins.

Interfaces between the relatively low voltage bi-polar TTL systems and electromechanical devices such as relays or displays operating at considerably higher voltage, and possibly in connection with MOS type arrays also capable of working at higher voltages, form a further group of applications particularly suited to thick film hybrid assemblies. Another aspect of the interface problem is that of noise immunity. Here again thick film circuits allow the selection of optimal components, and so can provide superior performance, compared with that of exclusively integrated circuit systems.

Thick films may also have advantages at the other end of the range in low power logic systems. Here the improved isolation obtained by mounting on a ceramic substrate can much reduce the power requirements, as compared with integrated circuit logic with its relatively low resistance 'intercomponent' isolation system involving back-biased pn diodes.[48]

Specialised applications where hybrid microcircuits may have further advantages in providing more accurately designable characteristics include: interconversion between analogue and digital systems, high speed clock drives, voltage and level shifters, and paper/card readers.[8,26,34]

A further special form of complex thick film circuit now becoming of increasing importance is the so called Multilayer System, in which the complexity (often of a high order) is topological rather than functional. In these circuits a number of conductor planes or patterns are printed over each other on a common substrate, but with intervening insulating layers. Through-holes or 'vias' in the insulating layers allow contacting at various levels of the several conductor networks to provide a highly flexible interconnection system. They are particularly suitable for high component density circuits such as those employing MSI and LSI, semiconductor unpackaged chips in logic circuits such as adders, and shift registers or MOS/bipolar memories.[36–43]

It will be clear from the above that thick film hybrid microcircuits have a very wide range of application. They are particularly suited to providing mother board interconnect systems for semiconductor active devices which are a vital interface if the full potential of semiconductor integrated circuits is to be realised, particularly MSI and higher complexity circuits.

Thick film hybrid circuits are capable of providing a precision in passive component values and performance still well beyond that at present foreseeable for integrated semiconductor technology; by addition of discrete components they allow the use of almost any of the very wide range of active devices now readily available, so that the best device for each circuit function can be chosen. Thus it is possible by utilising the full flexibility of the thick film approach to build circuits precisely tailored to a specific performance requirement.

REFERENCES

[1] Fahley, W. A., 'A Standardised Hybrid Microcircuit Design Cycle', *Proceedings IEEE Electronic Component Conference*, 503–513. (1970).

[2] Szekely, G. S., 'Classification and Technology for I.C. Technology', *Microelectronics*, 24–30. (February 1968).

[3] Beale, J. R. A., 'Microelectronics', *J. Physics D. Series* 2, **1**, (10), 1229–1240. (1968).

[4] Fosco, P. J., 'Thick Film R.F. Circuitry', *Proceedings IEEE Electronic Component Conference*, 289–297. (1969).

[5] Kirby, P. L., 'The Future Role of Film Circuits in Microelectronics', *Microelectronics*, 17–22. (July 1969).

[6] Cunningham, J. E., 'Microelectronics and Microwaves', *Electronic Design News*, 61–68. (October 1969).

[7] Wilkin, G. A., 'Thick Film Transistors', *Proceedings Conference on Thick Film Technology, Imperial College, London*, 93–98. (1968).

[8] Schwartz, P., 'For Design Flexibility go Hybrid', *Electronic Design*, 80–83. (July 1970).

[9] Thun, R. E., 'Thick Films or Thin', *IEEE Spectrum*, **6,** (10), 73–79. (1969).

[10] Nolan, R. W., 'Thick Films in Automotive Equipment', *Electronic Components*, 711–713. (June 1969).

[11] Denda, S. S., 'Hybrid Audio P.A.', *Proceedings IEEE Electronic Component Conference*, 236–242. (1970).

[12] Colwell, W. D., 'Reliability of Thick Film Modules in Auto Equipment', *I.E.E.E. International Convention Digest*, 442. (1971).

[13] Dyson, G., 'Is there a Place for Hybrids in Your Products?', *Canadian Electronic Engineer*, **15,** (2), 32–35. (February 1971).

[14] Hamer, D. W., 'Economics of Thick Film Hybrid Microcircuit Production', *Proceedings IEEE Electronic Components Conference*, 576–584. (1970).

[15] Asama, K., 'Development of Thick Film Hybrid Techniques in Fujitsu', *Fujitsu Science and Technology Journal, Japan*, **7,** (1), 101–130. (March 1971).

[16] Bigler, R. R., 'Equipment Designers Approach to Packaging of Hybrid Microcircuits', *Proceedings IEEE Electronic Component Conference*, 12–15. (1970).

[17] Gundotera, V. K., 'Important Aspects of In-House Thick Film Facility', *International Hybrid Microcircuit Symposium*, California, USA, 7.3.1–7.3.12. (1970).

[18] Engel, M., 'Hybrid Thick Film Chroma Demodulator and Colour Difference Amplifier,' *IEEE Transactions on Parts, Materials and Packaging,* **PMP 5,** (2), 117–122. (June 1969).

[19] Ottaviano, 'Thick Film Techniques and Design Criteria for Space Vehicle Applications', *Record of 8th International Electronic Circuit Packaging Symposium, Wescon,* 2.3.1. (1967).

[20] Furhoven, T., 'Hybrid Integrated Circuit P. A. for Transducers', *Proceedings of Semiconductor Integrated Circuit Conference.* (February 1971).

[21] Richardson, R. D., 'Problems in Packaging Thick Film Circuits', *Ibid.*

[22] Williams, J., 'Problems Associated with Hybrid Thick Film Assembly', *Ibid.*

[23] Matcovich, T. J., 'Interconnections in Hybrid Circuits', *Proceedings IEEE Conference on Integrated Circuit Development,* 240. (March 1971).

[24] Graff, W. S., 'Evolution from Hybrid to Monolithic Technology at IBM', *Ibid,* 636.

[25] Leyshon, W. E., 'An Overview of Hybrid Integrated Circuit Reliability', *Digest IEEE International Convention,* 606. (1971).

[26] Straub, J., 'Reliability of Hybrid Microcircuits in Use Today', *Proceedings IEEE Electronic Component Conference,* 16–27. (1970).

[27] Methven, B. S., 'Design, Construction and Performance of Thick Film Trimming Potentiometers', *Proceedings of International Hybrid Microcircuit Symposium, Imperial College, London,* 21–27. (1968).

[28] Weight, B., 'Miniature Adjustment Potentiometers', *Electronic Engineer,* 33–26. (August 1969).

[29] Johnston, S. A., 'Trimming Potentiometers', *Proceedings Electronics Conference* (*IEEE*) 185–194. (1969).

[30] McKie, L., 'Cermet Trimming Potentiometers', *Design Electronics,* 30–35. (August 1969).

[31] Harland, G. E., 'Hybrid Voltage Regulator for Automotive Applications', *Proceedings International Symposium on Hybrid Micro-Electronics,* 361–368. (1969).

[32] Glaser, A. B., 'A New Transformerless Modulator and its Hybrid Microcircuit Implementation', *Ibid,* 369–378.

[33] Miura, N., 'High Power Hybrid Integrated Circuit on Insulated Metal Substrate', *Ibid,* 379–396.

[34] Jung, W. G. 'Hybrid One-shot Logic Input and Bipolar Output', *Electronic Engineering,* **30,** (4), 66. (April 1971).

[35] Zook, G. E., 'Low Cost Thick Film R. F. Preamplifier', *Proceedings International Hybrid Microcircuit Conference,* 407–410, (1969).

[36] Loasby, R. G., 'Aspects of Multilayered Thick Film Hybrids', *Solid State Technology,* **14,** (5), 33–37. (May 1971).

[37] Close, A. D., 'Multilayer Substrates using Thin and Thick Film Techniques', *Ibid,* 38–42.

[38] Gioia, J. C., 'Multilayering for a Low Cost Hermetic Thick Film Substrate Package', *Ibid,* 43–46.

[39] Cox, J. J., 'Multilayer Ceramic Packaging', *Proceedings IEEE Conference on Integrated Circuit Development,* New York, 562. (1971).

[40] Keys, L. K., 'Fabrication of Multilayer Thick Film Microelectronic Circuits', *Proceedings International Hybrid Microcircuits Symposium,* **6.4,** 1–10. (1970).

[41] Theobald P. R., 'Multilayer Composite Substrates for Thick Film Hybrid Microcircuits', *Ibid,* 447. (1969).

[42] Rossman, M., 'Evaluation—Testing of Thick Film Multilayer Interconnection Systems', *Ibid,* **6.3,** 1–9. (1970).

[43] Ilgenfritz, R. W., 'Interconnection and Packaging of Thick Film Microcircuits', *Ibid,* 122–126. (1970).

[44] Venradt, J., 'Automatic Testing of Hybrid Circuits', *Instrument and Control Systems*, **44,** (4), 124. (April 1971).

[45] Gledhill, B., 'It pays to be Discrete', *Electronic Engineering*, 33. (May 1971).

[46] Barnwell, P., 'Thick Film Inductor at VHF', *Electronic Components*, 14. (January 1970).

[47] Corkhill, J., 'Properties of Thick Film Inductors", *Ibid*, 593–602. (May 1969).

[48] Nichols, 'Contribution of Thick Film Resistors to Logic Circuit Reliability', *Electronic Components*, 289. (March 1970).

[49] Platz, E. F., 'Solid Logic Computer Circuits', *Microelectronics and Reliability*, 55–59. (February 1969).

[50] Weinberg, E. M., 'A Thick Film Hybrid Precision Digital to Analogue Converter', *Proceedings International Hybrid Microelectronics Symposium*, (California, USA), 2.4.1–2.4.7. (1970),

[51] Bugbee, T. H., 'Opto-Hybrid Integrated Circuit Photosensitive Devices', *Proceedings of Solid State Sensors Symposium*, (Minneapolis, USA), 38–42. (June 1970).

[52] Schulz, W., 'A Major Step in Power Hybrids', *Proceedings 21st Conference in Electronic Components* (*IEEE*), 159–162. (1971).

[53] Shevkopylas, G. B., 'Matrix Capacitor for Trimming of Thick Film Hybrid Quartz Crystals Oscillators', *Radio, Electronics and Communications*, Soyst, (USA,) 13, 5. (May 1970).

[54] Perna, V. J., 'High Bit Rate Hybrid Circuits and Components as Seen from the Frequency Domain', *Proceedings International Hybrid Microcircuits Symposium*, (California, USA). 2.2.1–2.2.2. (1970).

[55] Reissing, T. C., "An Overview of Todays Thick Film Technology', *Proceedings of IEEE*, **59,** (10). 1448–1454. (October 1971).

[56] Mori, E., 'Headphone-type Stereo Radio, *RF–60*', *National Technical Reports* (*Japan*), **15,** (3), 263–274. (June 1969).

[57] Wyatt, R. E., 'The Design of Discrete Thick Film Resistors', *Component Technology*, **4,** (4), 10–12. (May 1970).

[58] Hoft, D., 'Hybrid Thick Film Techniques for HF–VHF Applications', *Electronics Progress*, **12,** (12), 10–17. (1969).

[59] Peterson, W. R., 'High Power Hybrids', *Design Electronics*, **8,** (1), 38–39. (October 1970).

[60] Engler, R. M., 'Design Considerations for a High Power Series Regulator', *Proceedings 21st Conference on Electronic Components* (*IEEE*), 163–175. (1971).

[61] Schneider, M. V., 'Hybrid Integrated Circuit Frequency Multiplier from 10 to 30 GHz.', *Proceedings IEEE*, **58,** 1402–1404. (1970).

[62] Otto, G., 'Thick Film Integrated Circuits in Telephony', *SEL Nachrichtentechnik*, **16,** (2), 38–44. (1968).

[63] Sarson, A. E., 'Hybrid Integration for Computer Circuits', *Proceedings International Conference on Microelectronics—Eastbourne*, 23–24. (June 1969).

[64] Kamp, F. S., 'A Hybrid 100 Watt Linear Audio Amplifier', *IEEE Transactions on Broadcast/TV Receivers*, BTR—**15,** (2), 142–150. (July 1969).

[65] Rouzier, M., 'MUR Subnanosecond Logic Circuit', *Proceedings International Conference on Advanced Microelectronics, Paris*, 930–939. (April 1970).

[66] Denda, S., 'Hybrid Audio Power Amplifier', *Proceedings IEEE Electronic Components Conference*, 236–242. (1970).

[67] McWilliams, D. A., 'The Research Challenge for Thick Film Technology' *Proceedings 21st Electronic Components Conference IEEE*, 434–435. (1971).

[68] Methven, B. S., 'Design, Construction and Performance of Thick Film Cermet Trimming Potentiometers', *Radio and Electronic Engineer*, **37,** (1), 17–21, (January 1969).

[69] Stinehelfer, H. E., 'Etched Thick Film Microwave Circuits', *NEREM Record*, 134–135. (November 1968).
[70] Greenwald, C., 'Printed Thick Film Microwave Integrated Circuit', *NEREM Record*, 140–141. (November 1968).
[71] Clar, P. L., 'Microwave Integrated Circuit Technology', *Electrotechnology*, **82,** (4), 37–42. (October 1968).
[72] Vergnolle C., 'Modular Conception of Hybrid Microwave Integrated Circuit Sub-assemblies', *Proceedings International Conference on Advanced Microelectronics, Paris,* 1015–1024. (April 1970).
[73] Hollenbeck, D. W., 'Multiloop Negative Feedback Active Filters using Thick Film Integrated Circuit Techniques', *Proceedings WESCON,* (San Francisco, USA), 4.4.1–4.4.5. (1969).
[74] Brayles, W. R., 'Improved R–C Active Filters using Integrated Circuit/Thick Film Technology', *Ibid.* 4.6.1–4.6.5.
[75] Smith, G. W., 'Voltage Regulator Capabilities using Hybrid Techniques', *Ibid.* 17.2.1–17.2.12.
[76] Langer, E., 'Typical Applications of Thick and Thin Films', *International Elecktronic Rundschau,* **23,** (6), 149–152. (1969).

3

Substrates

The material employed in the manufacture of the substrate plate and the physical, chemical and electrical characteristics of the final product exert a considerable influence on both the production processes and the characteristics of a thick film circuit. It is therefore necessary to consider the nature of the substrate material and manufacturing procedures in some detail.

The major requirements for substrates for thick film applications are:

1. A uniformly smooth surface texture, consistent, however, with permitting adequate adhesion of the fired thick film layers.
2. A minimum of distortion or bowing of the plate.
3. Capability of withstanding the normal firing temperatures, generally in the range of 500 °C to 1 000 °C.
4. Attainment of close tolerances on overall dimensions, including the positioning of holes.
5. High mechanical strength, high thermal conductivity, and good general electrical properties.
6. Chemical and physical compatibility with the normal commercial thick film conductor, resistor and dielectric glaze compositions.
7. Low cost in quantity production.

3.1. SUBSTRATE MATERIALS

Of the many ceramic materials which could be considered, for example, alumina, beryllia, magnesia, thoria, zirconia and ceramics

Table 3.1

CHARACTERISTICS OF CERAMIC MATERIALS

Material		85% Al_2O_3	96% Al_2O_3	99.5% Al_2O_3	99.5% *BeO*	*Magnesia* (MgO)	*Low Alkali Porcelain*	*Low loss steatite* ($MgO. SiO_2$)	*Forsterite* ($2MgO. SiO_2$)	*Cordierite* ($2MgO. Al_2O_3 5SiO_2$)	*Titania* (TiO_2)
Density	g/cm^3	3.4	3.75	3.9	2.9	3.4	2.6	2.7	2.9	2.3	4.0
Flexural strength	kg/cm^2 lb/in^2	2 800 40 000	3 500 49 000	3 300 47 000	2 500 35 000	1 500 21 000	1 500 21 000	1 700 24 000	1 400 20 000	1 050 15 000	2 000 28 000
Maximum working temperature	°C	1 400	1 700	1 900	2 500	1 100	1 200	1 000	1 000	—	—
Melting point	°C	1 910	1 950	2 000	2 570	2 800	1 655	1 460	1 890	—	2 130
Coefficient of thermal expansion $\times 10^{-6}$	20–200 °C 20–600 °C 20–1 000 °C	6.2 7.0 7.6	7.7 8.0 8.4	6.7 7.6 8.2	5.8 8.1 9.4	— 8.5 —	4.4 4.9 —	8.1 8.6 8.9	— 10.8 10.7	— 2.3 2.9	— 8.7 9.1
Coefficient of thermal conductivity at 25 °C	cal/cm. sec. °C	0.035	0.06	0.077	0.57	0.04	0.01	0.007	0.008	0.005	0.012
Thermal shock resistance	°C	230	200	180	205	—	200	175	—	330	—

Material		85% Al_2O_3	96% Al_2O_3	99.5% Al_2O_3	99.5% *BeO*	*Magnesia* (*MgO*)	*Low Alkali Porcelain*	*Low loss steatite* (*MgO SiO_2*)	*Forsterite* (2*MgO. SiO_2*	*Cordierite* (2*MgO. Al_2O_3.* (5*SiO_2*)	*Titania* (*TiO_2*)
Bulk resistivity	ohm. cm.	10^{14}	$>10^{14}$	$>10^{14}$	$>10^{16}$	$>10^{14}$	10^{14}	10^{14}	10^{16}	10^{14}	10^{16}
Surface resistivity	ohm/sq.	3×10^{11}	6×10^{11}	6×10^{11}	20×10^{11}	—	—	—	—	—	—
T_e (temperature at which bulk resistivity = 10^6 ohm.cm)	°C	850	1 000	1 000	1 240	—	350	275	—	—	—
Dielectric constant, κ	—	8.2	9.0	9.5	6.5	8.0	5.6	6.1	6.4	5.7	80
Dissipation factor, tan δ $\times 10^{-4}$	—	6	5–10	3–9	4	1	10–20	3–5	20	80	—
Breakdown strength	kV/mm. ac. rms.	9.2	9.5	10	9.8	10	18	12	20	12	—

(*Courtesy of 'Microelectronics'*)

based on these compounds, the first, alumina (Al_2O_3), offers by far the best combination of characteristics in relation to the above criteria. The properties of a number of the more common materials are compared in Table 3.1.

It is apparent that high purity alumina, that is alumina of greater than about 95% purity, exhibits a combination of the desired electrical, thermal and mechanical properties which is markedly superior to the remaining materials, with the one exception of beryllia (BeO).

The latter possesses the considerable advantage of a high value of thermal conductivity, (approximately nine times that of alumina), so enabling a circuit based on this material to dissipate substantially greater power. It is, however, rather weaker and relatively costly to produce in substrate form. Moreover it possesses highly toxic properties when in the finely powdered or vapour forms which would be produced respectively during machining and high temperature firing. Consequently, alumina of approximately 96% purity is currently used in the great majority of commercial thick film microcircuit applications. In spite of its drawbacks, beryllia is used in some specialised applications in which heat dissipation is of major importance.

Ceramic compositions based on barium titanate are sometimes used in specialised applications. Owing to their high dielectric constants, which may be in the region of 1 200–3 000, such materials are suitable for the production of capacitor networks in which the substrate plate *per se* constitutes the dielectric element. This application is discussed more fully in Chapter 6.

3.2. SUBSTRATE CHARACTERISTICS

Reverting to the standard material, 96% alumina, we can now discuss its characteristics in more detail in relation to the various criteria for thick film applications already outlined.

3.2.1 SURFACE FINISH

The surface characteristics of the ceramic substrate are of prime importance. A surface which is too smooth will result in poor adhesion of the fired thick film layers, while too rough a surface will lead to poor reproducibility of film thickness and, hence, to excessive variation in resulting circuit parameters.

In practice, the normal acceptable range for surface finish is from 20–40 μin, centre line average (CLA), the preferred figure being approximately 25 μin. Centre line average height, as measured from a trace of the surface profile, is defined as the arithmetical average value of the departure of the whole of the profile above and below its centre line throughout the prescribed meter cut-off, that is the wavelength cut-off corresponding with the sampling length in a plane substantially normal to the surface[1].

A number of methods and equipments are in use today for the evaluation of surface finish of ceramic substrates. These typically rely on the traverse over the ceramic surface of a probe, normally a diamond stylus, associated with a transducer which senses the vertical displacements of the probe during its travel. The transducer output, after amplification, is displayed as a meter reading of the average roughness value (CLA), while at the same time a trace of the surface undulations is provided by a chart recorder.

Different types of transducers to monitor the probe undulations have been used for such surface measurements. These include piezoelectric devices, inductive sensors, and more recently, semiconductor displacement transducers.

In one such test equipment, shown in Plate 3.1, a diamond stylus associated with a solid state displacement transducer and motorised drive traverses the substrate at a speed of 0.5 — 10 mm/min. The CLA value is read from a meter and the surface contour of the ceramic, in relation to an optically flat reference surface incorporated in the transducer head, is traced by a pen recorder. This depicts the general surface character and any local distortions or defects.

The same equipment is normally used also for the thickness and surface texture determination of the printed conductive, resistive and dielectric circuit layers.

Figure 3.1 shows the surface profile trace of an alumina substrate having a CLA value of the order of 25 μin.

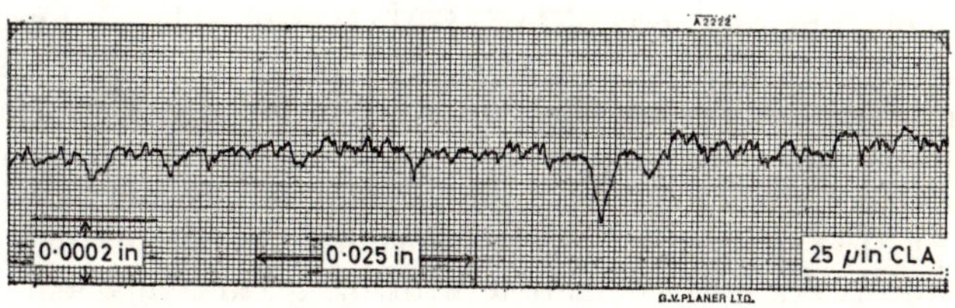

Fig. 3.1. Surface Profile Trace of Alumina Substrate (G. V. Planer)

3.2.2. DIMENSIONS

For general thick film applications, dimensional tolerances on length, width and hole positioning are typically of the order of ± 0.003–0.004 in. Substrate sizes range in current normal practice from about ½ in² to 4½ in². The substrate shape is produced by punching at the 'green' ceramic stage, so permitting a variety of shapes, where required with holes and notches, to be readily made; Plate 3.2 illustrates a number of typical substrates of different configurations.

Substrate thickness normally ranges from 0.020 in–0.040 in with 0.025 in being perhaps the most commonly used size; the thickness tolerances are usually set at between $\pm$ 0.001 in and $\pm$ 0.004 in. The extent to which substrates may be bowed depends largely on the overall size of the plate and typical values here are within about 0.004–0.005 in/in.

It is important to set close tolerances on substrate length, width and thickness, as well as hole position and degree of bowing, in order to maintain precision in the subsequent printing operation.

Excessive dimensional tolerances on the straight edges of the substrate will affect accurate registration of the substrate with the printing screen. Distortion of the substrate may effect the 'snap-off' distance, which is the distance between the natural position of the screen and the substrate, as well as the printing force between the screen and substrate. As a result of these factors, excessive bowing can cause considerable inconsistency in, for example, fired resistor characteristics. The relationship between these variables is discussed in greater depth in Chapter 7.

3.3. COMPATIBILITY WITH THICK FILM COMPOSITIONS

A further important factor is the compatibility of the substrate material, both in regard to physical and chemical characteristics, with the glaze compositions employed. A high purity alumina substrate comprises crystals of α-Al_2O_3 embedded in a glass/crystalline matrix normally composed of materials such as calcium and magnesium silicates. The latter are included as high temperature fluxes, and their proportion controls the sintering temperature of the ceramic and to a considerable extent, its ultimate mechanical and physical properties. It is, of course, important that these constituents are fully compatible with the glaze component of the thick film conductor, resistor or dielectric composition.

This question of compatibility will, of course, not normally concern the user. The manufacturer of the glaze composition will have aimed to develop his product to ensure the necessary compatibility with the standard alumina substrate material of approximately 96% purity.

The manufacturer of the usual glaze compositions will also

have ensured that the expansion coefficients of the fully fired glazes are as closely matched as possible to that of the alumina substrate material, since poor expansion matching would result in strains being set up. In the case of resistor films, these would tend to produce high temperature coefficients of resistance. In the worst case, crazing, or even loss of adhesion, of the films, whether resistive, conductive or dielectric, would occur during temperature cycling of the circuit. To avoid such adverse effects the expansion coefficients are normally designed so that the fired glaze is in compression with respect to the substrate.

3.4 SUBSTRATE MANUFACTURE

Since the surface texture, as well as many of the other characteristics of the substrate, depend upon the method of manufacture of the substrate, it is of interest to consider this subject briefly.[2,3] Two main methods are available, namely dry pressing and ribbon casting of the ceramic dough or slip prior to firing.

The former method can be used only for substrates in which the cross section to thickness ratio is relatively small, that is less than about 20:1. If this ratio is to be exceeded, it will be necessary to compress the ceramic dough to a thickness which is greater than that required, and subsequently to grind the fired ceramic down to the specified dimension.

The latter process will, however, tend to remove large crystals from the body of the substrate, and will also expose any natural voids within the body. This can well result in the formation of cavities as large as 0.002 in across and 0.001 in deep; in order to produce a satisfactory surface by grinding, very fine grain alumina powder must therefore be used in the original ceramic dough. The additional requirement for the use of a grinding operation will, of course, add substantially to the cost of the substrate.

The drawbacks of the dry pressing method are virtually overcome by the ribbon casting technique. In this process, alumina powder of fine particle size, chosen to give the desired surface finish, is suspended in a liquid carrier/binder medium to form a ceramic slip. The latter is then cast onto a glass carrier or other support presenting a smooth surface, the thickness of the layer being regulated by means of a 'doctor' blade. The material is then conveyed on a moving belt through an oven in which it is dried until sufficiently plastic to be handled.

The substrate configuration, including holes if required, is stamped out at this stage, and the 'green' substrates so produced are transferred to flat ceramic plates or batts which act as their support during the firing process. Special gas fired kilns are employed and firing schedules are chosen to give maximum densification of the ceramic whilst minimising distortion which could be caused by unduly rapid elimination of moisture or binder residues. Depending on the purity of the alumina used, sintering will normally be carried out at between 1 450 and 1 800 °C. After sintering, the substrate will remain unaffected in surface finish, size and structure by subsequent firings of thick film compositions.

The attainment of accurately controlled dimensional tolerances and consistent physical characteristics is, of course, closely linked with the casting and sintering process. In practice, the precision with which mechanical tolerances can be held varies from manufacturer to manufacturer, and the ultimate price of the substrate depends to an appreciable extent on these factors.

Composite multilayer alumina substrates, incorporating a buried noble metal conductor plane, have been developed for applications in which high component densities necessitate an increased number of electrical connections per unit area between components. The substrates are produced by screen printing a palladium interconnection layer on a green ceramic sheet. This is covered by a second green ceramic sheet, the composite then being compressed and fired at a temperature of about 1 500 °C. Connection to the buried conductor layer is made through holes, or 'vias', punched in the upper ceramic layer before assembling the composite structure.[4]

Multi-substrate plates (also known as 'Snapstrates') have prescored grooves in one or both surfaces, enabling them to be simply broken down to produce individual units. The potential advantage of this technique is that the composite plate, comprising a number of identical unit substrates, can be printed in one operation employing a multi-image screen, with consequent saving in cost. However, this advantage is to some extent offset by increased printing problems resulting from the greater variations in dimensional tolerances over the larger substrate areas.

REFERENCES

1 BS 1132:1961. Centre Line Average Height Method for the Assessment of Surface Texture, *British Standards Institution.*

2 Richards, G., 'Alumina Substrates in Thick Film Technology', *Electronic Materials,* 12–14. (July 1968).

3 Waterfield, B. C., 'Alumina Substrates for Thick Film Circuits.' *Microelectronics and Reliability.* **7,** No. 2, 117–119. (May 1968).

4 Theobald, P. R., Davies, P. M. and Bailey, J. T., 'Multilayer Composite Substrates for Thick Film Hybrid Microcircuits.' *Proceedings of Hybrid Microelectronics Symposium, Dallas,* 447–454. (1969).

4

Conductor Patterns

Metallic conductor patterns are used in thick film microcircuitry to provide low resistance circuit interconnections, resistor and capacitor terminations, and external connector bonding pads. They also provide bonding 'lands' for discrete capacitors and semiconductor devices, and metallic areas for subsequent hermetic sealing of protective lids. Various metals are used to make up the conductor composition, typical examples being gold, platinum–gold, palladium–gold, palladium–platinum–gold, palladium–silver, platinum–silver or silver.

When selecting a conductor composition for a particular purpose, the following factors, which are discussed in more detail in the following paragraphs, must be considered:

1. Ultimate solderability, (ease with which the solder wets the film) and resistance to leaching by the solder,
2. adhesion of the fired films,
3. suitability for wire and semiconductor chip bonding,
4. compatibility with resistor and dielectric films,
5. line definition attainable,
6. resistance to ageing effects on re-firing,
7. migration and other effects under the influence of d.c. fields.

4.1 COMPOSITION AND PRODUCTION

Thick film conductor compositions normally comprise the following constituents:

1. A metallic conductive component comprising one or more precious metals in finely divided powder form,

2. a bonding agent comprising a finely divided glass frit,
3. an organic suspension medium,
4. an organic diluent.

The compositions are produced by ball milling the constituents separately to submicron particle size, blending them in the requisite proportions, and then dispersing the mixture in the organic suspension medium and diluent to form a paste of the desired viscosity. In the case of a conductor composition, the metallic components may form up to 80% of the total weight of the composition whereas the glass content may be as low as 5% by weight. The highest total solids contents (85–90%) are found in those compositions which are designed for the printing of fine conductor lines, for instance, those having a line width of less than about 0.010 in.

The rheology of the composition is dictated by the subsequent printing conditions, such as the mesh of the screen, the required print thickness and line width, and by the necessity to maintain reproducible printing characteristics over a long period of time. These factors are discussed in greater detail in Chapter 7.

The firing cycle for the conductor composition after deposition on to the substrate consists of a warm up period during which the

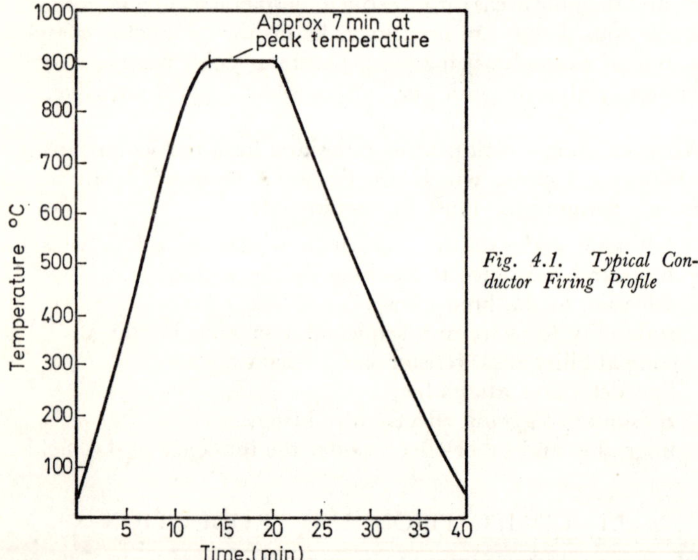

Fig. 4.1. Typical Conductor Firing Profile

organic suspension medium and diluents are eliminated by volatilisation or by burning off to form carbon dioxide, a period at peak firing temperature when the glass bonding constituent melts and sintering of the metallic particles takes place, followed by a cooling period to anneal the film. A typical firing profile is illustrated in Figure 4.1.

Firing is carried out using a specially designed conveyor furnace, the temperature/time profile of which is adjusted to conform to the firing schedule appropriate to the particular glaze composition used. A uniform air flow is maintained through the furnace in a direction opposite to that of the advancing substrates. This is to ensure that the volatile organic materials are removed from the furnace and do not come into contact with the glaze films at the elevated firing temperature.

An adequate air flow to ensure complete elimination of organic constituents is of considerable importance in practice, since failure in this respect can result in serious degradation in the properties of the resulting films. The characteristics of the conductor composition depend critically upon the peak firing temperature, which is normally in the range of 750–1 000 °C.

4.2 SELECTION AND CHARACTERISTICS OF STANDARD CONDUCTOR MATERIALS

We can now consider more fully the various factors determining the characteristics, and hence the selection of thick film conductor compositions for various practical applications.

4.2.1 CONDUCTORS FOR SOLDERING APPLICATIONS

For the majority of applications, the fired conductor should be suitable for soldering to permit the solder attachment of external connections, chip capacitors and semiconductor devices, or to obtain a general reduction in circuit interconnection resistance.

In the latter case, tinning the conductor by immersion of the substrate in molten solder will give a reduction in surface resistivity to about one tenth of the as-fired value. This operation is best carried out in a wave or flow soldering machine. Naturally only the solderable conductor areas of the type described will become tinned, while other conductors, as well as resistors, dielectrics and protective glaze areas will not be wetted by the solder. A circuit having tinned interconnection areas is illustrated in Plate 4.1.

Solderable conductor compositions can be tinned directly with conventional tin-lead solders preferably containing a small proportion of silver. Such solderable compositions include platinum–gold, palladium–gold and palladium–silver. A range of compositions having different metal alloy contents is available within each of these types, and cover a wide range of performance characteristics.

In selecting the conductor composition to meet the technical and economic requirements of the particular soldering application, the following aspects have to be borne in mind:

1. Solderability and/or bondability.
2. Adhesion.
3. Processing conditions.
4. Compatibility with thick film resistor and/or dielectric compositions.
5. Line resolution.
6. Conductivity.

The solderability of the film and its resistance to solder leaching, that is, to solution of the metal component in the molten solder, can vary considerably from composition to composition. Resistance to solder leaching largely depends on the metal system; of those referred to, palladium–gold alloys have the greatest and palladium–silver the least resistance in this respect.

Leaching occurs as a result of the ready solubility of silver and gold in molten solder such as 60/40 tin–lead composition. The rate of solution can be limited by minimising both the temperature at which soldering takes place and the time during which the conductor is in contact with the molten solder. On the other hand, these conditions have, of course, to be compatible with satisfactory tinning and adhesion of the components to be soldered.

Solder leaching can also be minimised by the use of tin–lead type solders containing a proportion of silver usually of the order of 2%, but sometimes up to 9%. The inclusion of silver in the solder effectively inhibits uptake of silver or gold from the deposited films. The silver content also affects the melting point of the solder; for example, commercially used compositions of 2% silver/62% tin/36% lead and 9% silver/57% tin/34% lead melt at 179 °C, and in the range 183–250 °C, respectively, whereas the standard 60% tin/40% lead solder melts at 188 °C.

Resistance to leaching is normally tested by measuring the period during which a standard circular conductor pattern can be maintained in molten solder before the first signs of conductor loss at the edge of the pattern are observed. Tests carried out, for example using 63% tin/37% lead solder at 215 °C, 4% silver/96% tin solder at 265 °C, and 10% tin/90% lead solder at 340 °C typically show 2–3 minutes resistance to leaching in the first case, ½–1 minute in the second, and up to 2 minutes in the third.

Initial solder acceptance, that is ease of wetting, is tested by measuring the time required for complete wetting of the standard conductor pattern during the solder/temperature tests described above. Typically, complete wetting is obtained during a 5 second dip; however solderability can be adversely affected by repeated heat cycles. Hence the wetting tests should be carried out only after the conductor pattern has undergone the complete cycle of conductor, resistor and dielectric firing treatments involved in the production process.

The adhesion to the substrate of a conductor intended for subsequent soldering is extremely important, especially where the joint may be stressed, as during normal handling of soldered discrete components. Initial adhesion figures of the order of 20–30 lb/in width in a 90° peel test are obtained with many of the conductor systems available. However some may show a reduction in the degree of adhesion as a result of subsequent firings of resistor and other circuit component films. Some reduction in adhesion may also occur as a result of long term ageing when the circuit is held at a raised temperature for a prolonged period.[1] In general the platinum–gold conductors show the least loss of adhesion during subsequent processing and ageing.

A number of compositions are available for the production of solderable fine-line conductors. The latter can be classed as those having track widths and track clearances of less than 0.010 in, the lower limit attainable in current practice being about 0.002 in. The solderable conductor compositions usually comprise palladium–silver, platinum–silver, platinum–gold, or platinum–palladium–gold constituents. For fine-line printing, the inks are compounded to give a thixotropic paste of high viscosity and high solids content (85–90%). The printing of this type of material requires special screens and precision techniques, and is fully discussed in Chapter 7.

The resistance of the fired films must, of course, be as low as possible, and is typically of the order of 5–20 Ω/in for a 0.005 in wide track of normal thickness. This value can be further reduced by subsequent tinning of the conductor pattern with solder. In the case of high silver content tracks, the conductivity may, on the other hand, decrease as a result of dissolution of silver in the solder.

Adhesion of the film is particularly critical in the case of fine-line conductors where the area of contact with the substrate is much reduced. Special attention has been paid by manufacturers of metallising paints to this aspect, as well as to the prevention of microscopic fissures in the fired films. Whilst the latter are relatively unimportant in conductors of standard dimensions, their presence in fine-line patterns can cause a complete break in a circuit path, which usually cannot be bridged, even by subsequent tinning.

Conductor compositions used for interconnecting thick film

resistors and capacitors must be fully compatible with the resistive and dielectric glazes. Poor compatibility will, in general, be apparent by interaction between the glassy phases of the respective layers at the firing temperature. This causes deterioration in the component characteristics, as well as crazing of the layers and loss of adhesion. In general the manufacturer of glazes will ensure that compositions within a particular range of products will be mutually compatible. Care must obviously be taken, however, where glaze compounds from different sources are to be used in the same circuit.

Some glaze manufacturers have, in fact, designed their thick film composition ranges so that the conductor layers are first printed and fired at the highest temperature employed in the production process. Subsequent layers are then printed and fired consecutively at decreasing temperatures. For this purpose the glassy phases present in the compositions are designed to have successively lower melting points, and the expansion coefficients of the fired films are matched as closely as possible.

Other glaze manufacturers have, on the other hand, designed completely compatible ranges of conductor, resistor and dielectric compositions which are suitable either for sequential firing or, if desired, for co-firing.[2,3] In the latter case, as many as six layers comprising conductors, resistors, crossovers and capacitors are printed individually on the substrate, each being dried in turn before the next is applied. The compositions are then co-fired in one operation, that is at the same firing temperature, to produce the different circuit elements.

The conditions used for firing the conductor compositions, as well as those for any subsequent firing of resistor or dielectric layers, strongly influence the properties of the ultimate conductor pattern. Adhesion and solderability are particularly sensitive to any processing variations. The optimum peak firing temperature of the conductor varies with the particular composition used, and the firing profile is generally selected to develop the highest degree of adhesion.

The ease of solder wetting is also a function of the peak firing temperature, and in the case of some compositions, this property is extremely sensitive to subsequent re-firing processes.

Degradation in the characteristics of the fired conductor can also result from other causes, for example, contamination from active fluxes, excessive exposure of the layers to molten solder, or prolonged heating.

4.2.2 CONDUCTORS FOR DIE AND WIRE BONDING APPLICATIONS

A second class of commercial conductor compositions is designed for the attachment of semiconductor devices by eutectic die bonding, thermocompression wire bonding and ultrasonic wire bonding techniques, rather than by soldering. The metallic constituent in this type of composition may be either pure gold or a high gold content alloy. Gold being readily soluble in the normal molten solders, these compositions are naturally unsuitable for soldering applications. However gold is compatible with the bonding of gold or aluminium lead wires as used in conventional semiconductor devices; gold–silicon and gold–germanium eutectic alloys are moreover readily formed, so making possible the attachment of silicon or germanium devices by eutectic bonding. These subjects are dealt with in more detail in Chapter 9, in the context of the devices to be attached.

As in the case of the solderable compositions, it is important to consider: line resolution, conductivity, processing conditions, wire and die bonding ability, and compatibility with thick film resistor and dielectric compositions, when selecting the most suitable composition for a given application.

Line resolution and conductivity are possibly even more important in this context than in the case of the solderable compositions. Line widths and spacing may well be required here to be compatible for example, with, beam lead devices having gold tape leads of widths as small as 0.002 in on 0.005 in centres. Whilst fine-line gold conductor compositions having high solids contents are currently available, the realisation of line widths of this order requires special techniques of precision printing (more fully covered in Chapter 7). Alternatively, line widths of the order of 0.002 in can be produced by photoetching thick film gold conductor layers.[4]

The conductivity of pure gold films is considerably higher than that of the solderable conductors, making them suitable for the production of fine lines in which resistivity can become of considerable importance. For example, a gold conductor 0.005 in wide and about 0.0005 in thick will have a typical resistance of the order of 2 ohms per inch of track. However, conductivity will of course vary with print thickness and it will be of importance to maintain close control over this factor in practice. For this purpose a regular check in production with a film thickness test instrument is desirable. When considering the resistance per inch of fine line conductors, it should be borne in mind that, although the solderable compositions have initially high resistance compared with that of the gold type, this can subsequently be reduced by dip-tinning in solder.

As in the case of the solderable compositions, the gold conductor materials must be compatible with the thick film resistor and dielectric compositions, and similar considerations are again applicable. Likewise, it is necessary to ensure that the composition selected is compatible with the subsequent process steps in the fabrication of the microcircuit.

Gold conductor compositions develop the optimum combination of properties when fired at about 950 °C, but subsequent processing steps can adversely affect the fired film properties. Gold conductors, for example, are not generally suitable for terminating the palladium–silver type of resistive composition. In circuits incorporating these two types, it will therefore be necessary to print and fire an additional conductor pattern of a fully compatible composition.

4.2.3 CONDUCTORS FOR THICK FILM INDUCTOR APPLICATIONS

Thick film inductors for use at frequencies of the order of 100 MHz have been produced using both standard palladium–silver and gold conductor compositions, both of which types are compatible with most other thick film materials.[5–7] To obtain printed spiral inductors having high Q values, it is necessary to employ, for a given physical size, maximum conductor widths, and to reduce skin effects by building up the print thickness to about 0.0018 in by multiple screening. The requisite low resistance is obtained by using low resistivity gold formulations. Alternatively, relatively high resistivity palladium-silver compositions are employed, the resistance of the inductor spiral subsequently being reduced by dip-solder tinning. Utilising such techniques and materials, Q values of 150 at approximately 120 MHz can be obtained.

4.2.4. SUMMARY OF CHARACTERISTICS

Typical characteristics of the four main metallurgical classes of standard conductor compositions currently available from one supplier (E. I. du Pont de Nemours) are summarised in Table 4.1. There are, of course, some variations in the specific properties of the various compositions available within each main class, as is the case in similar commercial compositions available from other suppliers.

4.2.5 COST

The cost of the conductor material is largely determined by the type and content of precious metal in the composition, and consequently the platinum–gold compositions are generally the most, and the palladium–silver the least, expensive. As a rough indication of the relative costs, it can be said that the prices per troy ounce of typical compositions of platinum–gold, gold, palladium–gold and

Table 4.1

COMPARATIVE CHARACTERISTICS OF CONDUCTOR TYPES

		Au	*Pt/Au*	*Pd/Au*	*Pd/Ag*
Optimum Firing Temp. °C		850	850	850	850
Resistivity, ohms/square		0.005–0.010	0.08–0.10	0.04–0.10	0.010–0.030
Solderability		nonsolderable	excellent	excellent	excellent
Bondability	wire	excellent	good	good	good
	die	excellent	good	good	non-bondable
Resistor Compatibility	Pd/Ag	No	Yes	Yes	Yes
	RuO_2	Yes	Yes	Yes	Yes
Line resolution, mils		5–15	5–15	5–15	5–15
Cost		High	Highest	High	Lowest

(*Courtesy, E. I. du Pont de Nemours.*)

palladium–silver are in the approximate ratios of 10:8:7:3, respectively. Whilst the actual cost of the conductor compositions may range from about £8 to £40 per troy ounce, the quantity used in a typical microcircuit in continuous production will be relatively small. On the basis of 1 troy ounce of conductor composition giving a coverage of 400 in^2 of ultimate conductor area,[8] the cost per 1 in × 1 in substrate, assuming an average type of circuit pattern, will be of the order of a few pence only.

4.3 COMPOSITIONS FOR CIRCUIT ASSEMBLY AND PACKAGING

Various conductive compositions are available for the assembly and packaging of circuits. These include compositions which, after firing, yield films which are suitable for brazing employing alloys with melting points in the region of 325–850 °C. Also solder or braze compositions having no glass content and which, after deposition on a fired conductor layer and subsequent heating, melt to yield a tinned surface.

A number of metallisation compositions for the attachment of components by brazing are available for hybrid circuit packaging. For this purpose Kovar or nickel plated Kovar leads, and Kovar or metallised ceramic covers can be attached to corresponding areas metallised with the brazeable compositions. Such materials, incorporating platinum or platinum–gold powders together with a glassy phase, are screened and fired as in the case of the standard types of conductor compositions. They have, however, to be highly resistant to dissolution of the conductor material in molten braze at elevated temperatures in the hydrogen/nitrogen atmosphere used for brazing. Table 4.2 lists a range of brazing alloys suitable for use with such compositions, together with the brazing temperature range for each alloy.

In assembling hybrid circuits utilising these compositions, one method is to position the lead frames or lids with braze alloy preforms of appropriate composition and configuration on the metallised substrate with suitable jigs to maintain the component positions. The assemblies are then placed in a furnace having a reducing atmosphere (typically 15% hydrogen/85% nitrogen) and the temperature is raised rapidly to the brazing temperature. This is held for the minimum period consistent with adequate flow of the brazing alloy. The temperature is then reduced relatively slowly to ambient.

Ranges of printable solder and braze compositions are also available. These comprise finely divided metal and alloy powders dispersed in a thixotropic organic vehicle, together with an organic reducing agent or flux of the non-activated, non-corrosive type. Such compositions are generally used to tin thick film microcircuits, semiconductor packages and microelectronic components such as

Table 4.2

BRAZING ALLOYS

Braze Alloy	*Brazing Temperature Range.* °C
88% gold–12% germanium	450–600
20% gold–80% tin	325–400
72% silver–28% copper	800–840
gold–silver–germanium	600–700

capacitors and resistors. Application is normally by screen printing, although brushing and 'dotting' from a tube can also be used. After deposition, the solder paste is reflowed by heating for about 10–15 seconds at a temperature between 20–50 °C above that of the melting point of the alloy. Table 4.3 lists some types of solder alloys commonly available, together with the melting point or liquidus.

Tin and lead based solders are typically used for the attachment of leads and passive components, and for the bonding of active devices, either in prepackaged form or as solder flip chips. The gold-tin solders are normally used for the hermetic sealing of packages with gold plated Kovar or metallised ceramic lids. They are also used for the attachment of passive devices or semiconductor chips at lower temperatures than those necessary for eutectic die-bonding. When employing these solders to tin thick film conductors it is necessary to consider, as already discussed, the resistance of the latter to solder leaching and to high temperature ageing effects.

Solder or braze compositions are used when it is necessary to deposit these selectively rather than to coat the entire conductive pattern, or where the effects of non-uniform solder deposition, which may occur with normal dip or wave soldering techniques,

Table 4.3

SOLDER MATERIALS

Type	*Melting point or Liquidus, °C*
100% tin	232
95% tin–5% silver	221
95% tin–5% antimony	240
60% tin–40% lead	193
10% tin–90% lead	300
80% gold–20% tin	280

are to be avoided. They also provide an effective alternative to the use of solder preforms, which tend to be difficult to handle and may not always be available in the desired shape or alloy composition.

These applications are more fully considered in the context of hybrid microcircuit assembly and packaging in Chapters 9 and 11 respectively.

4.4 OTHER CONDUCTIVE COMPOSITIONS

Other conductive compositions which are available from various suppliers include silver or copper glass frit based types, and gold or silver resin based types.

A wide range of glass based silver compositions are available having general characteristics similar to those of the standard precious metal types already discussed. They are not generally used, however, in high stability hybrid microcircuitry. This is as a result of the well-known tendency for silver to migrate under the influence of d.c. fields, and is liable therefore to make pure silver conductors incompatible with resistive films. They also have a relatively poor solder leach resistance, and there is a tendency for silver to tarnish in atmospheres containing, for example, traces of sulphur.

These disadvantages are, nevertheless, effectively overcome by inclusion of palladium in the composition, as in the case of the low cost palladium–silver conductor compositions. The use of silver compositions is, therefore, limited to such specialised applications as the production of electrodes for ceramic capacitors and the like.

A low cost approach to thick film conductor composition production is afforded by the use of glass frit compositions incorporating copper in place of silver or other precious metals. The cost per troy ounce of a copper based conductive composition is approximately one tenth that of a typical palladium–silver composition.

One copper based conductive ink having a fired resistivity in the region of 0.015 to 0.040 ohm per square is claimed to possess good substrate adherence, good solderability and to be suitable for normal printing techniques.[9] It is, however, necessary to fire the ink in an inert atmosphere such as nitrogen, in which the oxygen impurity content does not exceed 0.01%. At this partial pressure of oxygen there is little tendency for the copper to oxidise at the recommended peak firing temperature of 850 °C. It is, however, important that some oxygen is initially present to allow for burn-out of the temporary binder. Close control and monitoring of the firing atmosphere is therefore required in this case.

A wide range of conductive compositions comprising gold or silver with a resin base, typically of the epoxy type, is also available. Many are suitable for screen printing, whilst others are compounded for application by brushing, or by 'dotting' from a tube or with a needle.

Curing is carried out by heating at a relatively low temperature, normally in the range of 100–250 °C depending on the particular type of resin constituent. Also, depending upon the nature of the latter, maximum heat resistance of the ultimate film will normally be in the range of 125–300 °C.

Solderability depends both on the metal and the resin components. Some silver compositions are solderable using a 60/40 tin–lead solder containing 2–9% silver. They have a resin activated flux, and use a short heating cycle.

Applications of gold and silver epoxy resin based compositions in hybrid microcircuitry are generally limited to the attachment of capacitor chips, active devices and heat sinks.

REFERENCES

[1] Anjard, R. P., 'Thick Film Conductor Adhesion Testing,' *Microelectronics and Reliability*, Vol. 10, 269–275 (August 1971).

[2] *Thick Film Inks Data Sheet TFG/201*, Johnson Matthey Metals Ltd.

[3] *EMCA Data Sheets*, Electro Materials Corp. of America.

[4] Keys, L. K., Francis, F. J., Russo, A. J. and Herring S., 'Photoetching and Screen Printing of Conductor Patterns for Face-Down-Bonded Devices,' *Proceedings, IEEE 20th Electronic Components Conference*, 87–91. (May 1970).

[5] Corkhill, J. R. and Mullins, D. R., 'The Properties of Thick Film Inductors,' *Electronic Components*, (May 1969).

[6] Barnwell, P. G., 'Thick Film Inductors at VHF,' *Electronic Components*, 681–684. (June 1970).

[7] Barnwell, P. G., 'Thick Film Inductors and Their Use in Tuned Circuits at VHF', *Electronic Components*, 749–751, (9th July 1971).

[8] *Thick Film Microcircuitry Handbook*, Data Sheet C-2, Du Pont de Nemours.

[9] *Chemical Product Data Sheet 9100–V*, General Electric Co.

5

Resistor Patterns

The resistive glaze compositions available today for thick film microcircuitry applications offer a wide spectrum of characteristics, ranging from the low cost, general purpose to the high stability, precision types. The latter afford resistor stability characteristics approaching those of vacuum evaporated, thin film resistors at a cost which is intermediate between that of the thin film, and of the general purpose thick film types. Their characteristics are, in general, comparable with those of the chemically produced tin oxide film resistors, and are markedly superior to those of carbon composition resistors.

The compositions typically comprise: a resistive component, a glassy phase, which on melting during firing acts as a binder, an organic suspension medium and an organic diluent.

The electrical characteristics of the fired resistor depend, of course, mainly on the resistive component, which may comprise one of the following:

1. Palladium–palladium oxide–silver compositions,
2. mixtures of precious metals other than palladium and silver,
3. ruthenium oxide,
4. thallium oxide,
5. Indium oxide,
6. Tin oxide,
7. Tungsten–tungsten carbides,
8. Tantalum–tantalum nitride.

Depending on the resistive component, compositions are available having fired sheet resistivities of from 1 $\Omega/\square$ to 5 M$\Omega/\square$. It should be noted that, since the resistance of a conductor of rectangular cross-section is proportional to its length and inversely proportional to its area of cross-section (i.e. width × thickness), the resistance of a square of film of any size remains constant, provided that the film has constant thickness and composition. Sheet or surface resistivity, expressed as ohms per square, therefore affords a convenient comparison for films of approximately constant thickness.

Temperature coefficients range from + 500 ppm/°C to about − 800 ppm/°C, with several classes of materials giving coefficients within ± 100 ppm/°C over a wide range of resistivities. Long-term thermal and load stability may vary from several per cent down to the order of 0.1% or less, measured over a period of thousands of hours. This wide divergence in stability characteristics depends in the main on whether the resistive component undergoes a chemical reaction during the firing process making it susceptible to further chemical change in use, especially at elevated temperatures, or whether it is chemically inert and does not undergo such changes. The palladium–silver type of composition is typical of the former class, whilst the ruthenium oxide type is representative of the latter.

The following sections deal in detail with each class in the context of its chemical and electrical characteristics.

5.1 PALLADIUM–SILVER COMPOSITIONS

The earliest thick film resistor compositions were based on palladium–silver[1] and in fact this type is still in widespread use as a general purpose, low cost, resistive material. The compositions comprise palladium oxide and/or palladium metal, silver metal and glass frit powders dispersed in an organic medium. Chemical changes occur during the firing process and strict control of firing conditions is therefore necessary to develop consistently the desired resistor characteristics.

Firing, carried out in an oxidising atmosphere, results in the formation of crystalline phases which have been identified by X-ray diffraction studies as palladium oxide and palladium–silver alloy.[2] These together with thermo-gravimetric studies[3] have indicated that the following reaction occurs.

$$\mathrm{Pd} \xrightarrow[>330\,^{\circ}\mathrm{C}]{\mathrm{O}} \mathrm{PdO} \xrightarrow[>520\,^{\circ}\mathrm{C}]{\mathrm{Ag}} \mathrm{PdAg} + \mathrm{PdO}$$

According to the results of the gravimetric studies, a progressive increase in weight and resistivity occurs up to about 600 °C due to oxidation of the palladium. A reduction in weight and resistivity then takes place at higher temperatures, due to reduction of palladium oxide. However, neither reaction proceeds to completion, and moreover the observed changes are not entirely compatible at temperatures above 400 °C.

Since resistivity increases, particularly on prolonged heating, with no corresponding increase in weight, it is concluded that protection of some of the palladium and palladium oxide particles by the sintered frit and possibly by the palladium–silver alloy occurs. This inhibits further chemical change and reduces electrical continuity.

Thermal probe measurements indicate that the conduction process is controlled by the palladium oxide phase, rather than by the palladium–silver alloy phase.[4] The latter, which may contain 65–70% silver for typical compositions fired at 750 °C, tends to modify the conductivity of the system by improving the conductive linkages between the palladium oxide particles. The resistor is essentially a p type semiconductor and both positive and negative temperature coefficients of resistance (TCR) can result.

The magnitude of the TCR depends on the relative proportions of the non conducting glass and conducting phases, and on the palladium oxide–metal ratio in the fired film. In addition however, TCRs (as well as resistivity) depend on the nature of the substrate material, in particular, on the difference in expansion coefficients of the substrate and the glass matrix.

TCR values can vary by an order of magnitude, and resistivity by a factor of about two, for films deposited on different substrates.[4] For this reason, it is important to use the resistor compositions only in conjunction with the substrate material prescribed, normally alumina of approximately 96% purity.

The compositional equilibrium point reached is dependent upon the firing conditions. In order to attain reproducible characteristics it is necessary to maintain close control over the firing profile, and in particular, the time at peak temperature.

Control of atmosphere, and especially moisture content at a constant level, is also important, since hydrogen may be formed by reaction between water and carbon resulting from the decomposition of the volatile organic materials. Both products will then, by virtue of their reducing action, inhibit formation of PdO and so affect the equilibrium point.

The ambient atmosphere should not contain organic solvents (for example, from degreasing baths), or other impurities likely to result in the formation of carbon monoxide which would act as a reducing agent.

The equilibrium point attained during the firing process will still be susceptible to some extent to change by oxidation of the resistor at elevated operational temperatures, with or without electrical load. It is also affected by reduction by atmospheres such as forming gas (H_2/N_2 mixture) which may be employed when sealing the circuits in protective cans.

The fired films are also susceptible to changes induced by contact with organic resins, particularly those incorporating reducing type hardeners. Such oxidation or reduction effects will be relatively small compared with those occurring at the elevated firing temperature. Nevertheless they may in some cases result in changes as high as several per cent of resistor values.

Changes due to electrical load or thermal exposure will be due to two effects. Initially, the temperature rise will increase the rate at which unoxidised palladium in the film combines with oxygen in the air to give an increase in resistance, the rate falling as the available palladium is oxidised. During the initial operational period of the resistor this reaction is, however, counterbalanced by a reduction in the contact resistance between metal or metal oxide particles by interdiffusion. In the early stages of load-life this reaction will tend to result in a decrease in resistance.

Thus, in compositions having a high metal content, a decrease in resistance occurs initially, whilst in high resistivity formulations having a lower metal content, a positive resistance drift is observed initially. During the latter stages of ageing, the mechanism of oxidation of palladium metal to palladium oxide predominates, and a positive drift in resistance values will occur

In order to reduce resistor drift, the palladium–silver type of resistor must, therefore, normally be protected with a glaze overlayer, which is fired at about 500 °C.

In normal production practice the sensitivity of resistors to firing conditions results in a spread from batch to batch of fired resistance of about $\pm$ 20%. In order to produce resistors to a closer tolerance, it is therefore necessary to 'trim' each fired element individually, using for example an air abrasive or laser trimming system, as discussed in Chapter 10. Trimming by such means is carried out after deposition and firing of the protective overglaze layer. Since this results in the exposure of one edge of the resistor track to the atmosphere along the trimmed length, the process may bring about a change in the stability of the element. This is, however, generally small. Differences in stability between trimmed and untrimmed resistor values amount to less than 0.1% measured over a period of 2 000 hours.

Table 5.1 summarises the temperature coefficient of resistance, noise and stability characteristics of a typical commercial palladium–silver resistor series.[5] Stability curves for the compositions[6] indicate that drift and load life are dependent on the firing temperature used in the production, whilst firing duration has a negligible effect.

Optimum stability appears to be achievable for firing temperatures in the region of 750 °C. Significant loss in stability during tests without electrical load is observed in films which have been fired at or below 730 °C. Resistors under electrical load normally exhibit relatively low drift rates, whilst exposure at 150 °C without load appears to accelerate the drift.

5.2 PRECIOUS METAL

In order to overcome the relative instability inherent in the palladium–palladium oxide–silver resistive system, composition ranges based on the noble metals and their compounds, which are free of palladium or silver, have been developed. Since the noble metals are chemically highly stable, they are less prone to oxidation or reduction effects than the palladium–silver materials.

Table 5.1

TYPICAL CHARACTERISTICS FOR PALLADIUM–SILVER TYPE RESISTIVE COMPOSITIONS

		7800 *Series*
Range	Ω/□	40–100 K
TCR ppm/°C	25–125 °C −55–25 °C	+ 250 → 0 + 50 → − 500
Noise, dB		− 22 → + 20
Stability % Δ R 1 000 hours.		No load, 150 °C < 0.5% → < 3% 40–60 W/sq in, 70 °C < 0.5% → < 1%

(*Courtesy, E. I. du Pont de Nemours*)

Two ranges have been described,[7,8] and some typical characteristics are given in Table 5.2.

The precious metal compositions are fired in an oxidising atmosphere at a temperature in the region of 980–1 050 °C. If close control over printing and firing conditions is exercised, it is possible to maintain resistance tolerances of ± 10% consistently on a production basis. TCR tracking, that is tolerance on TCR values, is quoted as 5 ppm/°C for arrays of elements with similar resistance values on a single substrate.

The fired films are insensitive to oxidising and reducing atmospheres, and to the presence of conventional commercial resin encapsulants; they can accordingly be used without the addition of protective overglazes. In the case of one series of compositions of this type, stability on short term exposure to forming gas (5% H_2/95% N_2) at 400 °C is claimed to be below 0.25% resistance drift.[8] As a result of the stability of the precious metal resistive component, exposure of the latter by partial removal of the glaze constituent by air abrasive trimming, has no significant effect on the load stability of the films.

Table 5.2

TYPICAL CHARACTERISTICS – PRECIOUS METAL TYPE RESISTIVE COMPOSITIONS

	ESL 3800				
Range Ω/□	1–10 M				
	10	1 K	10 K	100 K	1 M
TCR ppm/°C −55° to 125 °C	+ 50 ± 100	0 ± 100	0 ± 50	0 ± 50	−50 ± 150
Noise, dB	−20	−15	−10	0 ± 5	+15 to 20
Voltage coeff. ppm/V	5 k Ω/□—negligible 100–200 k Ω/□ 10–20 1 000 k Ω/□ 60–80				
Stability	200 °C, no load, 1 000 hours, ΔR, < 0.3% 24 °C, 25 W/sq. in., 1 000 hours, ΔR, < 0.2% After heat and voltage stabilisation, ΔR, 0.01–0.05%				

(*Courtesy, Electro-Science Laboratories Inc.*)

A further advantage offered by the precious metal resistor system is that, on account of the high firing temperature used, metallic terminals can be applied using compatible conductor compositions after firing of the resistor element. In this case, the resistor is first printed and fired, and is then contacted in a subsequent operation. In this way the formation of a step at the resistor/conductor interface, which could tend to affect the uniformity of printing of resistor tracks, is avoided. Post-termination also allows the entire conductor area to be available for soldering or lead attachment, so permitting the design of smaller overall contact lands. Post-firing will cause a small shift in sheet resistivity and TCR, depending upon the temperature used to fire the termination.

Co-firing of the resistor and conductor is also possible with precious metal compositions, provided that specific compatible conductors only are used. However, somewhat larger deviations in resistance and TCR than those normally resulting from pre- and post-firing may result in some cases.

5.3 RUTHENIUM OXIDE COMPOSITIONS

A number of resistive compositions employ ruthenium dioxide (RuO_2) as the resistive component. This material has, for a metal oxide, an unusually low resistivity, i.e. 5×10^{-5} Ω cm, so permitting its use without additional metallic particles to yield resistivities of practical interest for thick film circuits. Ruthenium oxide is highly stable, both towards oxidation and reduction at elevated temperatures and may be heated to about 1 000 °C in air without undergoing chemical change.

Resistors prepared from inks containing simple binary mixtures of ruthenium oxide powder and lead glass give films with a wide

Table 5.3

TYPICAL CHARACTERISTICS—RUTHENIUM OXIDE TYPE RESISTIVE COMPOSITIONS

	Birox	*1100 Series*
Range Ω/□	10–1M	10–1M
TCR ppm/°C − 55° to 125 °C	± 100	< 250
VCR ppm/V/in.	< −20 (at 10 K Ω/□)	−20 (at 100 K Ω/□)
Noise, dB	100 Ω: −30; 10 K: −10; 100 K: 0	100 Ω: −30; 100 K: 0
Stability	30 W/in², 70 °C, 3 500 hours ΔR < 0.2%	30 W/in², 70 °C, 1 000 hours ΔR < 0.2%
	No load, 150 °C, 1 000 hours ΔR < 0.2%	No load, 150 °C, 1 000 hours ΔR < 0.3%

(*Courtesy, E. I. du Pont de Nemours*)

range of electrical resistance, good stability and relatively low values of TCR and electrical noise.[9] In practice a dopant is normally added to modify the lattice structure of the ruthenium oxide and so provide a means of controlling the sheet resistivity and TCR over a wide range of values. Niobium pentoxide has been found to be a suitable dopant, with silver powder as a useful addition.[10] The glaze component has a significant effect on the electrical properties, and it is therefore necessary to include a homogeneous glass phase which is stable during the melting and firing stages to ensure consistently low TCR values.

Since the resistive component is highly stable at elevated temperatures, no chemical interaction takes place during firing. The attainment of specific resistance properties is not, therefore, critically dependent on the firing profile and atmospheric control. Hence much greater variation in these conditions is permissible than in

Table 5.4

TYPICAL CHARACTERISTICS—ALLOYS UNLIMITED RUTHENIUM OXIDE TYPE RESISTIVE COMPOSITIONS

		Series A						
Range Ω/□		10–10 M						
		10	100	1 K	10 K	100 K	1 M	10 M
TCR	25° to 125 °C	+230	+40	+50	+50	−200	−350	−400
ppm/°C	−55° to 25 °C	+230	+40	+50	+50	−250	−425	−500
Noise, dB		−35	−35	−20	−5	+15	+25	+25
Stability		150 °C, no load, 1 000 hours, ΔR, 1% max. 0.25% av. 80 °C, 25 W/in², 1 000 hours, ΔR, 0.2%						

		Series B					
Range Ω/□		10–1 M					
TCR	25° to 125 °C	±200	±150	±150	±150	±250	±350
ppm/°C	−55° to 25 °C	±200	±150	±150	±150	±250	±350
Stability		100 °C, 112 W/in², 1 000 hours, ΔR 0.25% max. 0.15% mean 150 °C, 112 W/in², 1 000 hours, ΔR 0.40% max. 0.20% mean					

(*Courtesy, Plessey Inc.*)

the case of the reactive palladium–silver type of composition. However, to obtain optimum print-to-print and batch-to-batch reproducibility, the conditions of application and firing, having once been established, should be controlled as well as the equipment permits. Firing is generally carried out in air at temperatures within the range of 750–850 °C.

The inert nature of the ruthenium dioxide also renders the fired films relatively insensitive to subsequent processing steps. These involve reheating for device bonding, contact with forming gas at elevated temperatures when brazing Kovar leads or protective cans, contact with solder fluxes or resin encapsulants, etc.

Similarly the films also exhibit excellent long term thermal and load stability, and permit significant improvements in high power performance to be obtained. As a result of the high stability, no protective over-glaze is normally necessary. Likewise, exposure of the resistive component of the films during abrasive trimming has an insignificant effect on their long term stability.

The reproducibility of resistance values obtained in normal production employing ruthenium oxide resistor compositions is typically of the order of $\pm$ 10%. Tables 5.3 and 5.4 summarise typical properties of ruthenium oxide based resistive elements produced from paints from different commercial suppliers.

The long term stability offered by some commercial ruthenium oxide resistor compositions is very high. Resistance changes are generally of the order of 0.1–0.2% over the course of several thousand hours test. It is interesting to note that this order of stability is, in fact, approaching that obtained with conventional vacuum deposited thin films.

5.4. THALLIUM OXIDE COMPOSITIONS

Thallium oxide (Tl_2O_3), like ruthenium oxide, is another semi-metallic compound having a low intrinsic resistivity of about 5×10^{-4} ohm cm. It enables a wide range of sheet resistivities to be obtained, when dispersed in a suitable glass matrix. Thallium oxide type resistive paints are available, specifically formulated either for use with 96% alumina substrates, or with substrates of higher expansion coefficients, such as forsterite, steatite or titanate ceramics.[11]

These compositions are distinguished from all other types, so far discussed, in that the firing temperature is relatively low, that is of the order of 570 °C, and firing can be carried out in an oxidising, inert or even reducing atmosphere. The resistivity of thallium oxide glaze films varies to some extent with peak firing temperature, but this effect is minimal where the latter is in the neighbourhood of 570 °C. Other electrical characteristics such as TCR, electrical noise, long term stability etc., vary only slightly, if at all, with firing temperature over the range of 550–620 °C.

The long term (1 000 hr) stability of thallium oxide films when loaded to 10 W/in^2 at 70 °C is of the order of 1%. TCRs are between + 50 and − 200 ppm/°C over the temperature range − 55° to + 150 °C. The use of a protective over-glaze is unnecessary in the case of high resistivity films. However, films having surface resistivities below about 1 000 $\Omega/\square$ require the use of a protective layer to attain similar characteristics, since less glassy phase is available in this case to provide adequate atmospheric protection. For this purpose, a silicone or low temperature fired glaze layer is suitable.

The thallium oxide glazes are, depending on their formulation, compatible either with palladium–silver conductors, or with gold constituent conductor compositions.

An important aspect of the thallium oxide glazes is their comparatively low cost, this being approximately half that of the conventional low cost palladium–silver and about $\frac{1}{8}$ that of the noble metal composition types.

Thallium oxide is a highly poisonous material, and in using these compositions, special precautions have to be taken to prevent inhalation of the dust or vapours, or accidental ingestion. Efficient exhaust systems have to be employed, both in conjuction with the firing furnaces and the resistor trimming equipment.

5.5 INDIUM OXIDE COMPOSITIONS

A resistive glaze composition using doped indium oxide (In_2O_3) as the resistive component has also found use in thick film circuit application.[12] The resistivity of the system is controlled by doping the indium oxide, by controlling the crystallite size of the indium oxide, and by varying the glass concentration.

Indium oxide is an n-type semiconductor, and its resistivity is affected by doping with ions having a valency which is higher or lower than 3. Pentavalent antimony or arsenic have both been used to lower the resistivity. The addition of 2% antimony, which on account of its ionic size provides a good fit in the indium oxide lattice, reduces the resistivity of indium oxide resistors by two orders of magnitude. Since antimony oxide has a high vapour pressure, some loss of dopant will occur during high temperature processing, and fired film properties will vary according to the firing schedule employed. For example, an increase in the time from 5 to 15 minutes at a peak firing temperature of 950 °C results in a threefold increase in resistivity.

The firing schedule normally used comprises a 60 minute cycle including 10 minutes at the peak temperature of 950 °C, firing being carried out in air.

Thermal drift is positive and is probably associated with oxidation of the indium oxide, resulting in an increase in the degree of stoichiometry and associated loss of carriers in the n-type system. Over 4 000 hours at 175 °C, resistance changes of up to about 2% are reported. Variations in resistance of unprotected resistors on load and humidity test are generally of the order of 1–2%. Such changes can be reduced by about 50% when the resistors are protected with a resin coating.

5.6 METAL CARBIDE AND NITRIDE GLAZES

Glass based resistive glazes comprising tungsten–tungsten carbide,[13,14] tantalum–tantalum nitride,[15] and titanium–titanium nitride[16] conductive components have been developed for the production particularly of discrete resistors, although they could naturally be used also in thick film circuits.

In the first type the resistive component comprises a mixture of 80% by weight tungsten carbide and 20% tungsten. These constituents are dispersed in a glassy phase, the ratio of glass to conductor controlling the resistivity of the fired film. Firing is carried out in an inert or a reducing atmosphere (nitrogen or nitrogen–hydrogen mixture respectively) at a temperature in the range 950–1 000 °C.

TCRs are within ± 250 ppm/°C, these values being predominantly positive in the lower resistive ranges, decreasing through zero and changing to negative as the conductor content is reduced and the resistance values are increased. In load stability tests carried out on $\frac{1}{4}$ W resistors at 70 °C, resistance changes after 1 000 hours were generally below 0·5%.

In the tantalum–tantalum nitride resistive system, the components in finely divided powder form are mixed in the weight ratio of 1:2, the resistivity range then being controlled by the amount of glaze incorporated. Firing is carried out in an inert nitrogen atmosphere at a temperature between 1 100° and 1 190 °C. The substrates employed are normally alumina or steatite.

Resistors are produced as described for the tungsten–tungsten carbide type; the TCRs here are below ± 100 ppm/°C and load life tests at 70 °C over 2 000 hours gave less than ± 0.45% resistance change in the case of the $\frac{1}{4}$ W elements tested.

5.7 TIN OXIDE GLAZE

Tin oxide has been used as the base of a glaze resistor material in the production of stable high value resistors in the range of 10^7–10^{12} Ω, mainly for use in current monitoring devices.[17]

The basic resistor component comprises stannic oxide doped with antimony. Conduction in undoped tin oxide films can occur since the lattice has a proportion of missing O^{2-} ions, the conduction mechanism being n-type. The electrical properties can therefore be controlled by introduction into the lattice of a known proportion of ions of different valency, such as Sb^{5+}, P^{5+}, In^{3+} and B^{3+}. In the case of antimony, the controlling valency mechanism is described by the equation

$$Sn^{4+}O_2^{2-} + Sb^{5+} \rightarrow Sn_{1-2x}^{4+}.Sn_x^{3+}.Sb_x^{5+}.O_2^{2-}$$

By this mechanism pentavalent antimony atoms replace some quadrivalent tin atoms, causing these to move to a lower valency state and releasing electrons into the conduction bands. The electrical conductivity is then proportional to the amount of antimony entering the crystal lattice. The maximum amount accommodated is between 7 and 10%, depending upon the method of production of the material, and compositions in this range exhibit minimum resistivity and minimum TCRs.[18]

Tin-antimony oxide powder for the high value resistive glaze applications has been prepared by oxidation of powdered tin-antimony alloys of the appropriate compositions. This can be carried out by feeding a continuous stream of the powdered alloy into a furnace held at 1 000 °C in which it is exposed to a stream of oxygen, so producing the oxide in a finely divided form. This material is subsequently mixed with a glaze constituent to produce the thick film composition. Resistors have been produced either by dipping the substrates into the composition, or screen printing, followed by firing at red heat.

Resistors of this type have been made with values in the range of 10^7–10^{12} Ω; the TCR is normally in the range of −1 000 to −15 000 ppm/°C, varying in a linear manner with temperature from −55 °C to 125 °C.[17]

5.8 COMPARISON OF RESISTOR COMPOSITIONS

Having discussed each type of resistor composition separately in some detail, we are now in a position to compare their relative merits in the manufacture of thick film circuits.

The most important property for comparison is the reactivity of the resistive component incorporated in the composition, since essentially this controls the ultimate characteristics of the fired film, whether in protected or unprotected form.

Of the commercially available resistor compositions, those based on palladium–silver are of the reactive type, whereas those based on precious metals, ruthenium oxide or thallium oxide are nonreactive.

Owing to the dependence of the palladium–silver type on an oxidation/reduction reaction occurring at the firing temperature, it is difficult to obtain resistor elements consistently within a tolerance better than ±15%; indeed in production practice, a more usual figure is ±20%. In the case of ruthenium oxide and precious metal compositions, since attainment of fired film characteristics is far less dependent upon close control of temperature and furnace atmosphere, it is possible to obtain consistent resistance values to within ±10% of the target value in production.

In the case of the reactive palladium–silver type, to attain optimum stability under usual environmental and electrical load conditions, it is generally necessary to apply a protective glaze coating over the resistors. This, on the other hand, is not normally required for the nonreactive ruthenium oxide and precious metal types. Similarly, thallium oxide resistors above 1 000 ohms per square do not require a protective glaze, while for the lower values, where a lower proportion of the glassy phase is present, such protection is again necessary.

Stability of the palladium–silver types under normal test conditions of load and temperature is generally of the order of 0.5 to 1 or 2% resistance change in the first 1 000 hours, whereas for the nonreactive types, the corresponding figure is only 0.1 to 0.5%. In general, the nonreactive materials are capable of dissipating a greater power per unit area of resistor film than are the reactive types, for a given resistance change. TCRs in the nonreactive type are typically lower than those in the reactive type. For some commercial materials of the former type the TCR can, in fact, be held to below ±100 ppm/°C over the entire resistance range (10 ohms to 1 megohm per square). However, the corresponding figure for the reactive type is typically of the order of ±200 to ±300 ppm/°C, with values rising to perhaps −500 ppm/°C at the higher end of the resistance range. Furthermore, in some cases electrical noise values for the nonreactive materials are below those in the reactive types.

The costs of the different types of resistor composition vary widely. In general, the noble metal types are the most expensive, the ruthenium oxide types being intermediate, and the palladium–silver types relatively inexpensive. The least expensive compositions on the other hand are the thallium oxide types.

As in the case of the conductor compositions, it should however be borne in mind that, under production conditions in which wastage can be kept to a minimum, the actual cost of the paint in a typical thick film circuit may be only a small fraction of the overall cost of the circuit element. This is apparent from the fact that the coverage afforded by 1 troy ounce of, for instance, a standard palladium–silver paint when printed through a 165 or 200 mesh screen may be as large as 410 and 570 in^2 respectively.[19]

REFERENCES

[1] Hoffman, L. C., 'Precision Glaze Resistors', *Bulletin, American Ceramic Society*, **42,** No. 9, 490–493, (1963).

[2] Melan, E. H. and Mones, A. H., 'The Glaze Resistor—Its Structure and Reliability', *Proceedings of Electronic Components Conference, Washington*, 76–85, (1964).

[3] Krahl, P. H. and Bogenschutz, A. F., 'Thermogravimetrische Untersuchung der Einbrennreaktion von Siebdruckwiderstandspasten', *Metall*, **22,** (10), 988, (1968).

[4] Melan, E. H., 'Stability of Palladium Oxide Resistive Glaze Films', *Microelectronics & Reliability*, **6,** 53–65, (1967).

[5] Technical Data Sheets, E. I. Du Pont de Nemours.

[6] Headley, R. C., 'Reliability Characteristics of Palladium–Silver Thick Film Resistors', *Proceedings, ISHM Symposium on Hybrid Microelectronics*, (1967).

[7] Stein, S. J., Garvin, J. B. and Vail, M., 'Thick Film Resistor Plates for High Performance Use', *Proceedings, Hybrid Microelectronics Symposium, Dallas*, (1969).

[8] Lane, G., 'Designing with High Reliability Thick Film Resistors'. *Proceedings, ISHM, Hybrid Microelectronics Symposium*, (1967).

[9] Angus, H. C. and Gainsbury, P. E., 'Glaze Resistors with Ruthenium Dioxide', *Electronic Components*, 84–88, (Jan. 1968).

[10] Iles, G. S., 'Ruthenium Resistor Glazes for Thick Film Circuits', *Radio and Electronic Engineer*, 299–304, (Nov. 1968).

[11] van Loan, P. R., 'A Thick Film Resistor Glaze of Precision Properties'. *Proceedings, IEEE Electronic Components Conference, Washington*, 285–288, (1969).

[12] Block, M. L. and Mones, A. H., 'Properties of Indium Oxide Glaze Resistors'. *Proceedings, IEEE Electronic Components Conference, Washington*, 191–196, (1966).

[13] Merz, K. M., Huang, C. and Murphy, R., 'Tungsten Carbide—Tungsten Resistive Glazes'. *Proceedings, IEEE Electronic Components Conference, Washington*, 1–7, (1966).

[14] Casey, H. B., 'A New High Quality, Low Cost Resistor', *Proceedings, IEEE, Electronic Components Conference, Washington*, 197–203, (1966).

[15] Merz, K. M. and Huang, C., 'Nitride—Metal Resistive Glazes,' *Proceedings, IEEE, Electronic Components Conference, Washington*, 292–298. (1968).

[16] Huang, C., 'Titanium Nitride/Titanium Resistive Glazes.' *Proceedings, IEEE, 20th Electronic Components Conference*, 34–39, (May, 1970).

[17] Dearden, J., 'Thick Film High Value Tin Oxide Glaze Resistors', *IERE/IEE/ISHM Proceedings of Conference on Thick Film Technology*, 53–60, (April, 1968).

[18] Phillips, L. S., 'Transparent Conductive Metal Oxide Films', *R. & D.*, No. 19, 46–49, (March, 1963).

[19] *Thick Film Handbook*, Data Sheet C–2, Du Pont de Nemours.

6

Printed Capacitors and Insulating Layers

Screen printed compositions comprising a dielectric material are employed in three basic applications: capacitor dielectric layers, insulant layers at resistor track crossover points, and in multilayer circuitry, protective overlayers for resistors and capacitors.

The dielectric compositions available conform to the general pattern already met in conductor and resistor compositions. That is, the appropriate powdered component, in this case the dielectric material, is dispersed with a glassy binder component in an organic suspension/diluent medium. The compositions are printed and fired using standard techniques.

6.1 CAPACITORS

Capacitors can be incorporated in hybrid circuits by screen printing and firing dielectric compositions, by utilising a high permittivity substrate as the capacitor dielectric and screen printing and firing on its surface the appropriate electrode pattern, or by attaching discrete chip devices.

Chip capacitors are, in practice, employed more widely than the screen printed types, since they can provide appreciably higher capacitance/unit area values than the latter. Moreover, they can be incorporated in the circuit by relatively simple soldering or bonding operations, while the screened type normally requires three screen printing and at least one firing operation, or the use of non standard high K substrate materials. In this Chapter, both screen printed and fired types, dielectric glaze and dielectric substrate, will be discussed. Chip devices will be described in Chapter 9.

6.1.1 DIELECTRIC GLAZE CAPACITORS

A thick film capacitor comprises a base electrode layer, a dielectric intermediate layer and an upper, or counter electrode layer. The dielectric compositions employed must, of course, be compatible with the standard types of conductor used for the interconnections and resistor terminations.

In production, the base electrode is usually first printed and fired in the normal way. The dielectric layer is then deposited, generally by printing and drying two separate layers, which are subsequently fired together. The double printing technique is employed to ensure freedom from pin holes, since the second layer will fill any holes or discontinuities in the first, so reducing the incidence of subsequent capacitor failure by voltage breakdown at pin hole areas. To complete the capacitor, the upper electrode layer is then printed and fired.

Alternatively, many proprietary dielectric compositions are suitable for co-firing with specified electrode compositions. In this case the base electrode, dielectric and upper electrode layers are successively printed and dried, the composite structure being then fired at the requisite temperature. In other cases, only the dielectric and upper electrode layers are suitable for co-firing.

Printed and fired capacitors can be adjusted to value in a number of ways:

1. Removal of a portion of the capacitor by air abrasion trimming; selective removal of part of the counter electrode, or removal of all three layers may here be employed.
2. Re-firing the capacitor under controlled conditions of time and temperature, since the permittivity is dependent on the firing conditions.
3. Employing a capacitor incorporating parallel trimming limbs representing small increments in capacitance; by cutting the limbs, the capacitance value may, depending on the configuration of the array, be adjusted, for example, from a 10% tolerance to within 1% of the required value[1].

The characteristics of the fired capacitor layers depend upon the nature of the dielectric material incorporated in the composition. In order to obtain an understanding of the limitations inherent in the use of screened capacitors, it is of interest to consider the subject in some detail.

Dielectrics in this context fall generally into the two classes of materials with low or high permittivity. The latter, so called high K type, is largely based on the ferroelectric ceramic, barium titanate, in which the permittivity at room temperature is approximately 1 200, reaching a peak value of 10 000–12 000 at the Curie point in the region of 120 °C.

In preparing a high permittivity composition the objective is, firstly to displace the Curie point, and hence the peak permittivity, towards room temperature. Secondly the relationship between permittivity and temperature is smoothed to permit the production of capacitors having relatively low temperature coefficients of capacitance. This may be achieved by adding, for example, strontium, calcium, tin and/or zirconium oxides to the barium titanate. Compositions based on (Ba, Sr, Ca (O)), (Ti, Sn, Zr (O_2)) have permittivities in the region of 1 000–3 000 and temperature coefficients of about ±5 000 ppm/°C in the range 20–100 °C.

Addition of lead oxide tends to increase the Curie temperature of barium titanate; however, when combined with compositions of the type (Ba, Pb)O, (Ti, Sn, Zr)O_2, very high permittivities, up to 25 000, and temperature coefficients of approximately ±20 000 ppm/°C, may be obtained in the region of room temperature.

Several low permittivity, non ferroelectric dielectric materials are alternatively available for use in dielectric compositions. These include magnesium titanate (K = 12–15), zinc titanate (K = 18–19), titanium oxide (K = 90–100), and calcium titanate (K = 160). The first two have temperature coefficients in the region of ±100 ppm/°C, whereas titanium oxide has a temperature coefficient of −800 ppm/°C. The availability of these materials has led to the development of temperature compensating capacitors (NPO types). The basic feature in this case is the addition of varying proportions of magnesium and/or zinc oxide to titanium oxide to give a composition which has a predictable and controlled temperature coefficient.

A number of proprietary thick film compositions incorporating dielectrics of the types discussed, and in which K ranges from a minimum of about 12 to a maximum of about 1 200 (measured at 1 kHz) are available. The maximum capacitance per unit volume obtainable using high K compositions is at present limited to about 30 000 pF cm^{-2} mil^{-1} (190 000 pF in^{-2} mil^{-1}). Dissipation factors (power factor) at this level of capacitance are of the order of 3.5–4%, decreasing to about 1–2% for compositions having values in the range of 50–800. Values of the breakdown strengths are typically in the region of 100–400 V d.c. mil^{-1} and insulation resistance figures are typically of the order of 10^{10} Ω cm^{-2} mil^{-1} when measured at voltages in the range of 10–100 V.

Fired high K ceramic capacitors are voltage, frequency and temperature dependent, the changes obtained being characteristic of the ferroelectric material incorporated in the composition, and therefore, similar to the changes observed in conventional ceramic capacitors.

Typical variations in capacitance and dissipation factor with d.c. and a.c. voltage, frequency and temperature for a high K dielectric are illustrated in Figures 6.1, 6.2 and 6.3 respectively. The permittivity obtained also depends on the peak firing temperature and Figure 6.4 illustrates this dependency for a typical composition having a nominal K of 500.

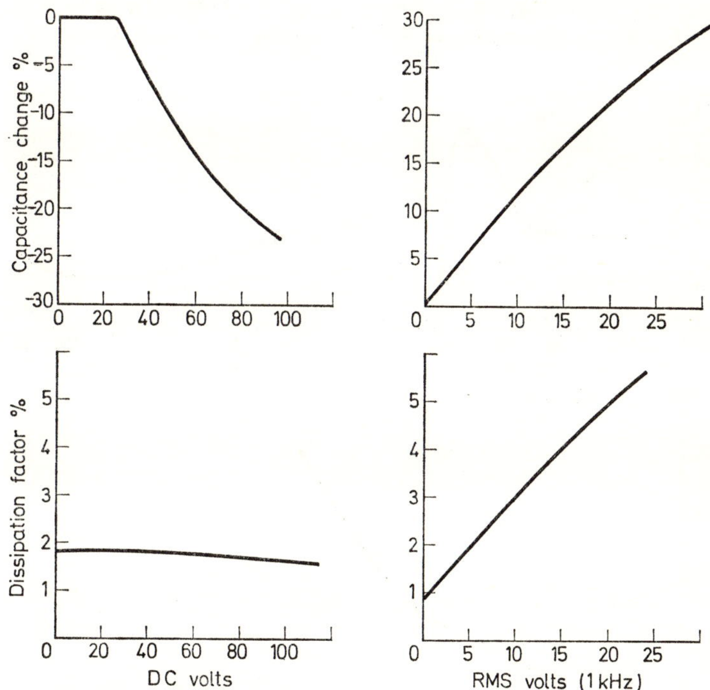

Fig. 6.1. Variation of Capacitance and Dissipation Factor with Voltage

High K capacitors exhibit an ageing effect which is attributable to realignment of the ferroelectric domains and is brought about by gradual relief of strain energy. The ageing effect is manifested as a slow reduction in capacitance and, for fired thick film capacitors, may be of the order of several per cent per decade hour when measured at 85 °C and 50 V. The change tends to be most rapid

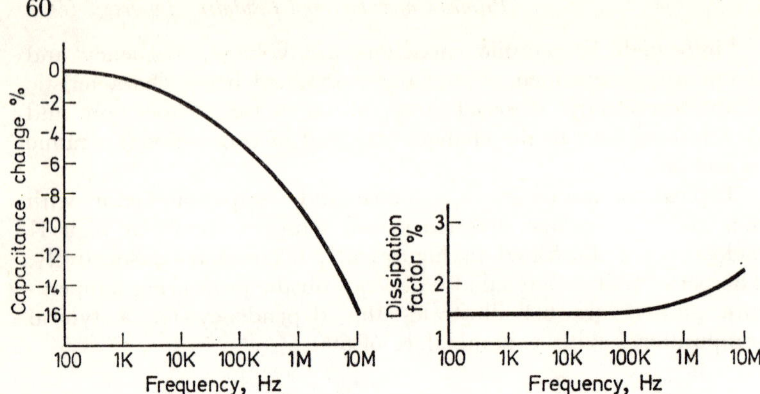

Fig. 6.2. Variation of Capacitance and Dissipation Factor with Frequency

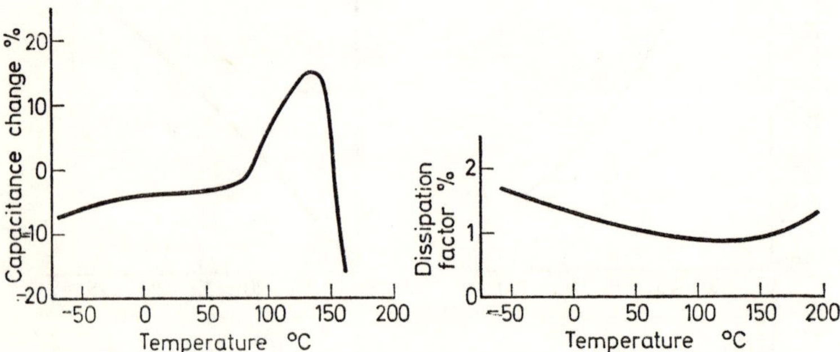

Fig. 6.3. Variation of Capacitance and Dissipating Factor with Temperature

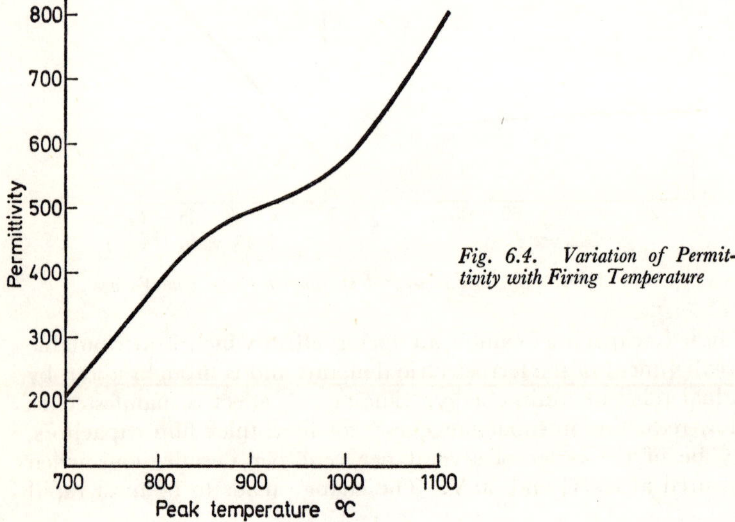

Fig. 6.4. Variation of Permittivity with Firing Temperature

in the period immediately after firing and some reduction in the overall change can be obtained by a heat stabilisation treatment[2], although a relatively slow ageing will still tend to occur thereafter.

Compositions are also available having a comparatively low dielectric constant in the range 12–20, but a correspondingly high Q (quality factor = dissipation factor^{-1}) of the order of 500–700 at 1 MHz. Such compositions yield capacitors having a higher degree of temperature stability, (typically about +1% linear change between 25–125 °C), than the high K compositions. The upper limit of capacitance attainable is however only about 300 pF cm^{-2} mil^{-1} (1 900 pF in^{-2} mil^{-1}).

The production of medium permittivity thick film capacitors from dielectrics comprising compounds of a ternary lead oxide–barium oxide–niobium pentoxide system has been reported[3]. Depending on the compositions of the materials, both ferroelectric and non ferroelectric dielectrics have been produced. The latter type exhibited a considerably higher degree of stability with time than the former, as would be expected as a result of the characteristic ageing effect exhibited by ferroelectric materials due to domain realignment.

Depending on dielectric composition, permittivities of thick film capacitors fired at an optimum temperature of 1 000 °C were in the range of 50–160 and dissipation factors were generally between 0.05–2.0. Both positive and negative temperature coefficients of capacitance have been obtained, the latter, up to about —1 000 ppm/°C, being associated with the highest permittivities.

6.1.2 DIELECTRIC SUBSTRATES

In the second approach to the production of screen printed capacitors the substrate *per se* constitutes the dielectric, and the electrodes are provided by printing and firing conductor layers.

The substrates employed for this purpose are normally of the high K barium titanate or doped zirconate–titanate type, and these are suitable for the production of both capacitor and resistor–capacitor networks.

Some proprietary conductor compositions are claimed to be directly compatible with such substrate materials, performing as well on these as on the standard alumina substrates, whereas others require modifications to optimise their fired characteristics.[4] In the case of resistors deposited on this type of substrate, the electrical characteristics produced differ markedly from those deposited on alumina substrates. This factor must therefore be considered at the design stage, and it will also be necessary to re-evaluate printing and firing conditions to optimise the characteristics of specific compositions for use with such special substrates.

Two types of electrode configuration can be employed, the parallel plate (double surface) and interdigitated (single surface) configurations, as illustrated in Figure 6.5 (a) and (b).

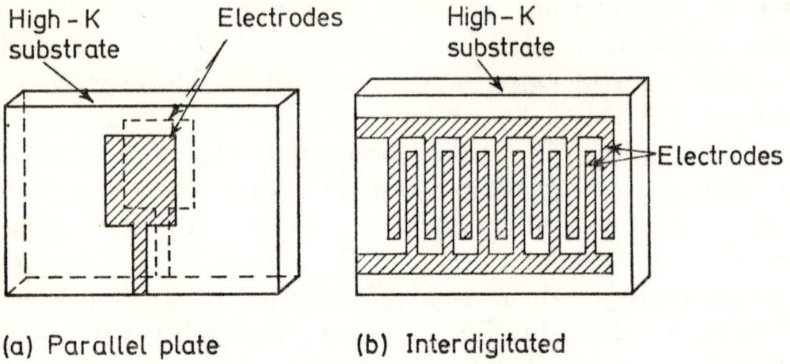

Fig. 6.5. Printed Capacitor Configurations

The interdigitated configuration has a number of advantages over the parallel plate configuration[1]:

1. Screening is restricted to one side of the substrate only.
2. No registration problems are involved, since the capacitor is produced in a single printing operation.
3. Terminals need be affixed to one surface of the substrate only.
4. The capacitor value is independent of the substrate thickness, and substrate dimensions can thus be selected on the basis of other factors such as mechanical strength, flatness and cost. Although significantly higher capacitance per unit area is theoretically possible in the parallel plate type if thin substrates are employed, in practice the values for both types are approximately equal in the case of substrates compatible with normal thick film processing, that is having a thickness of the order of 0.025–0.040 in.

The dielectric substrate approach offers the advantages that capacitor production is relatively simple. Also capacitors can be readily adjusted to value, for example by air abrasive trimming to remove a portion of the electrode limb. There are, however, also drawbacks to this approach. The range of dielectric materials available is limited, and those with relatively high permittivity yield capacitors having comparatively high power factors and high temperature coefficients. Moreover the dielectric substrates available are mechanically inferior to alumina and appear to be more difficult to manufacture to the high dimensional and flatness tolerances required in thick film practice. In addition, the dielectric substrates have lower thermal conductivity than alumina, and it is therefore necessary, if resistors are also incorporated on the substrates, to derate these by a factor of about 50%.

When using dielectric substrates, other components of the circuit may give rise to stray capacitances for which allowance must be made in the design. It has been found empirically,[1] that a circuit on the same side of the substrate as the capacitors increases the capacitance by about 5 to 10%, whereas a circuit on the opposite side of the capacitors increases the value by 20 to 25%.

6.2 DIELECTRIC GLAZES FOR CROSSOVER AND MULTILAYER CIRCUITRY

The increasing complexity of high density thick film circuitry has necessitated the development of multilayered circuits in order to deal with the complex interconnection requirements. Techniques and materials are now available to permit the production of as many as ten levels of circuitry. These are analagous to the earlier developed multilayer printed circuit board arrays but are smaller in size by a factor of about ten.

In this chapter the dielectric materials and relevant printing techniques will be discussed. In Chapter 9 multilayer thick film circuit arrays will also be considered in the context of hybrid circuits comprising attached discrete semiconductor and other chip components.

Three basic types of dielectric compositions are available, namely those incorporating glasses, recrystallising (or de-vitrifying) glasses, and glass/ceramic mixtures.

6.2.1 GLASSES

In the earliest approach to the problem, glass dielectric compositions were employed for the production of simple insulant layers at conductor crossover points, and for elements comprising two circuit layers separated by one insulant layer.

The glass dielectric compositions require a graduated firing schedule, the first conductor layer being fired at the highest temperature, the vitreous layer some 50–100 °C lower, and the upper conductor a further 50–100 °C lower than the intermediate vitreous layer. In this type of structure, the firing procedures employed are critical, since excessive flow of the glass will result if firing is carried out for too long or at an excessive temperature. This can cause lateral movement or 'swimming' of the upper conductors, resulting in distortion and partial loss of registration. Diffusion of the upper conductor into the lower layers may also occur, with consequent risk of short circuiting to the lower conductor layer.

Interconnection between the conductor layers is made through windows or 'vias' in the glass insulation layer, through which the next layer of conductor is subsequently printed.

6.2.2 RECRYSTALLISING GLASSES

A significant advance resulted from the development of crystallisable or devitrifying glass compositions which, in contrast to the vitreous materials, do not reflow on heating subsequent to the first firing.[5,6] During the latter process, the materials lose their vitreous characteristics and develop a crystalline structure. This prevents subsequent flow on reheating to temperatures well in excess of the original firing temperature. Although some flow of the glasses occurs during the first firing, dimensional stability is closely maintained during subsequent firing stages. This permits considerably increased yields in multilayer structures comprising many more layers than the three to which they would normally be limited by the use of vitreous insulants.

Since the dielectric, once fired, retains its dimensional stability on subsequent re-firing at the same temperature, the requirement for graduated firing temperatures is obviated. Co-firing of as many as nine alternating conductor and dielectric layers has been reported.[5]

Typical proprietary crystallisable glass dielectric compositions have: insulation resistance values greater than 10^{13} Ω at 100 V applied potential, dielectric breakdown strengths above 500 V (0.002 in thick layers), permittivity values in the range of 10–40 (1 kHz), dissipation factors below 1.5% (1 kHz), and Q values of 500–700 (1 MHz). After first firing at temperatures generally of the order of 850 °C, the recrystallisable glasses can be fired to at least 1 000 °C without appreciable loss of dimensional stability.

6.2.3 GLASS/CERAMIC MIXTURES

The advent of crystallisable glass compositions enabled considerable advances in multilayering to be made. Such materials nevertheless suffer from the disadvantage that the recrystallisation process involves phase changes, the firing process being therefore critically dependent on time and temperature. Although such glasses will not flow once they have been fired, a little movement will take place during the initial firing process with the result of some small loss in dimensional accuracy.

In order to overcome these disadvantages an approach has been developed in which the viscosity of the glass at normal firing temperatures is increased by the addition of a nonreactive ceramic component to reduce the amount of flow during the first firing.[7] In addition, gradual solution of the ceramic component in the glass on subsequent re-firing yields a further increase in viscosity. This in turn, further restricts both the glass mobility and reaction with the glass components of the metallising compositions. In practice the flow during firing of the glass component is typically restricted to between 0.001–0.002 in.

The development of high viscosity glass materials of this type is claimed to present a number of further advantages. Since the composition of the glass phase is not critical, any firing temperature may be specified. This can therefore be selected to be close to that for the metal compositions to be employed, and compatible insulants may hence be formed for different commercial metal pastes. In this way more freedom is also provided to optimise the values of physical properties such as permittivity and thermal expansion coefficient. Another advantage of the high viscosity glasses is that windows can be more readily provided.

In order to improve dimensional control and so increase yields when producing multilayer circuits employing such high viscosity glasses, a complementary printing technique has been developed.[7] This eliminates the so called 'ploughed field' effect which results in the thinning of the insulation where this is printed over raised lines, and which may cause crossover failure and loss of dimensional control.

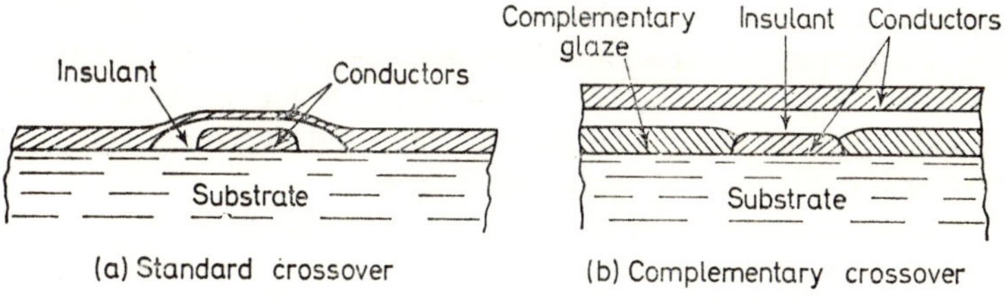

Fig. 6.6. Comparison of Standard and Complementary Crossovers

The complementary printing technique (illustrated in Figure 6.6) overcomes these difficulties by restoring surface flatness after each printing stage. This is carried out by printing the complement of the previous pattern, thus filling in the interspaces in the pattern.

In the case of a metal conductor layer a complement of dielectric insulant is used and vice versa. The complement of the metallisation pattern is normally arranged to be about 0.0002–0.0003 in thicker than the conductor film itself, so that the subsequent insulant layer deposited over the metal will be thickened. In order to reduce the demands made on the accuracy of registration of the printing screens, the complementary pattern is normally designed to provide a clearance around each conductor line of approximately 0.001 in. This gap is sufficiently small to be readily filled during the subsequent screening operation.

Typical yields obtained employing the complementary printing technique are claimed to be between 99.90 and 99.97%.

When printing through-holes, the yield is increased by first printing a 'picture frame' pattern around the hole area employing a highly viscous dielectric paste. A thick layer of a less viscous dielectric is then printed over the circuit outside the frames which act as walls. This prevents the glass from flowing into the holes during the subsequent firing process.

The requirements of ease of through-hole printability and freedom from pin holes are in opposition when formulating a paste incorporating high viscosity glasses. The former necessitates a reduction in the viscosity of the paste and the latter a reduction in the ceramic powder content of the loaded glass composition. Considerable compositional control is therefore necessary, the basic requirement being to formulate the paste so that a flow of 0.001–0.002 in occurs either on printing or firing.

6.3 ENCAPSULANT GLASSES

Encapsulant glaze compositions are employed to provide protection for printed resistors and capacitors from: environmental conditions such as high humidity or exposure to reducing atmospheres, reactive organic encapsulating materials, and overspray occurring during the air-abrasive trimming process.

Encapsulant compositions normally comprise vitreous materials whose melting points are considerably lower than those of the other types of dielectric glasses discussed. Such encapsulant compositions are printed in the normal manner, allowing an overlap of some 0.020 in around the resistor or capacitor and, after drying, are fired at temperatures of the order of 500 °C. Several proprietary ranges of encapsulant glasses are available; these include coloured materials incorporating some 2% of a metal oxide pigment.

REFERENCES

[1] Clough, W. L., 'Capacitors Compatible with Thick Film Circuit Technology.' *The Radio and Electronic Engineer*, 349–358, (Dec., 1968).

[2] Electro-Science Laboratories, Technical Bulletin.

[3] Edelman, F. H., 'Some Properties of Thick Film Niobate Capacitors', *Proceedings IEEE, 20th Electronic Components Conference*, 40–48, (May, 1970).

[4] 'Substrates for Thick Film Microcircuitry', E. I. du Pont, *Technical Publication*, 2, 8/68.

[5] Stein, S. J., 'Glass-Ceramic Glazes for Crossover and Multilayer Screened Circuitry', *Proceedings, IEEE, Electronic Components Conference*, Washington, 118–129, (May, 1968).

[6] Cox, J. J. and Hoffman, L. C., 'Multilayer Ceramic Wiring Structures for Hybrid L.S.I.', *Proceedings Inter NEPCON*, Brighton, (Oct., 1968).

[7] Loasby, R. G. 'The Technology of Multilayered Thick Film Circuit Arrays', *Microelectronics*, 12–16, (April, 1969).

7

Printing Procedures

The printing of thick film microcircuits is based on the long established technology of silk screen printing. The process depends essentially on the forcing of an ink through a stencil pattern comprising a fine mesh. It is widely used for such purposes as the production of display and decorative finishes, labelling of bottles and containers, and in electronics, in the production of printed circuit boards. For application to thick film microcircuitry, it has been necessary to develop the basic techniques, materials and machinery to an appreciably higher degree of refinement to achieve the required standard of precision and reproducibility in the printing.

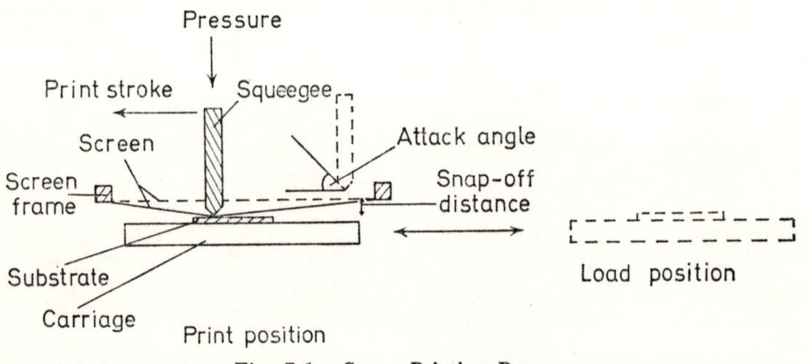

Fig. 7.1. Screen Printing Process

The screen printing process, as illustrated diagramatically in Figure 7.1, involves positioning of the ceramic substrate on a carriage, which is then moved beneath the screen so that the substrate is in accurate registration with the pattern on the screen; the latter is photographically defined so that open mesh areas in the screen correspond to the configuration to be printed, the remaining areas being blocked off.

In standard techniques of printing, the substrate is positioned a short distance beneath the screen, the clearance between the two generally being termed the break-away, or snap-off distance. A small quantity of the ink composition is dispensed onto the upper surface of the screen. A flexible wiper, the 'squeegee', then travels across the screen surface, deflecting it vertically to bring it into contact with the substrate, and forcing ink through the open mesh areas of the pattern. After the passage of the squeegee, the screen regains its original position through its natural tension, leaving the printed ink pattern on the substrate. The substrate carriage is then moved from beneath the screen, the substrate is replaced and the printing process repeated.

In the following sections, we shall consider the characteristics of screen printing machines as designed for microcircuit production applications, with particular reference to the various kinds of screens available and to the types of squeegee used. The factors controlling the precision and reproducibility of printed patterns obtained will in particular be reviewed in relation to the features of the printer.

7.1 THE SCREEN PRINTER

The basic functional components of a screen printer are:

1. The screen, together with its mounting framework,
2. The substrate carriage and feed system,
3. The squeegee mechanism and pressurising system,
4. The adjustment mechanism for the precise positioning of the screen relative to the substrate.

The screen and squeegee system may be distinguished in Plate 7.1 which illustrates a typical commercial printer.

7.1.1 SCREEN MOUNTING

The screen mounting is designed to clamp the screen frame rigidly and securely. For normal production applications, however, it must also permit rapid removal and replacement of screens, whilst retaining the location of the screen frame precisely and reproducibly to an accuracy of about 0.0005 in.

The screen mounting is provided with means for adjusting the screen in the X, Y and rotational sense to allow adjustment of the screen relative to the substrate, and to establish its position parallel with the substrate. Adjustments are normally made by micrometer screws, and screen-to-substrate registration can, in some commercial printers, be held to an accuracy of the order of 0.0003 in.

Facility for adjusting the clearance between the screen and substrate is also provided. This distance, generally between 0.01 in and 0.04 in for screens with a size of approximately 5 in $\times$ 5 in, is of considerable importance since it is one factor which controls both the amount of ink delivered through the screen and the precision of the printed pattern.

7.1.2 SUBSTRATE HOLDER AND CARRIAGE

The substrate holder comprises a platform having a recess in which the substrate is positioned. The latter is normally located by butting two adjacent edges firmly against datum surfaces in the recess. In addition, a vacuum may be applied through holes in its base to aid retention of the substrate in the recess.

Alignment of the screen with the substrate during setting up operations is facilitated if provision is made for both transmitted and reflected light in the substrate area. Cold illumination is necessary since any undue temperature rise would cause loss of registration accuracy due to expansion of adjacent metallic components.

The substrate holder is mounted on a moving carriage which travels from a loading position remote from the screen to the print position beneath the latter. The carriage movement must be both accurate, and a precise mechanism to ensure that the substrate to screen registration in the print position is repeatedly accurate to within about 0.001 in.

Most commercial printers incorporate carriages having either a reciprocating or rotary movement, the substrates here being moved into position beneath the fixed location screen. In one class of machine, however, the substrate holder remains stationary and the screen is mounted on an upper platen which moves vertically on precision ground die posts. The substrate is loaded into the holder and the upper platen is moved vertically down on the die posts in a similar manner to that of a precision metal stamping die press. It is claimed that, in this type of system, alignment between substrate and screen can be maintained to an accuracy of 0.0003 in over several million printing cycles.

7.1.3 SQUEEGEE MOUNTING

The squeegee mounting system incorporates a number of microcontrols to allow adjustment of the parallelism of the squeegee with the screen, the squeegee printing pressure, and the angle of attack of the squeegee; the latter is the angle between the squeegee blade printing surface and the screen (See Figure 7.1).

In all machines other than simple laboratory types, movement of the squeegee across the screen is automatic, with means of adjustment of the speed between about 0.1 and 10 in/second. In some machines the printing stroke is unidirectional, the squeegee remaining in contact with the screen only on the forward stroke, being then raised out of contact on the reverse stroke. In this case a flood blade is provided to return the excess ink to the starting point during the reverse stroke prior to the next printing cycle. In other machines, provision is made for two directional printing, the squeegee remaining in contact with the screen on both the advance and return movements.

It is essential to ensure smooth movement of the squeegee over the screen, in particular at very low speeds, avoiding erratic travel characterised by a juddering motion of the squeegee assembly. For this reason a damped hydraulic system, rather than a pneumatic drive system, is often used.

An automatic ink dispenser is incorporated in some printers. This may comprise a disposable syringe from which controlled amounts of paste are delivered to the squeegee through a plastic tube. In this manner, loss of volatile materials from the ink, and hence variations in viscosity, are avoided. However, in most cases, the ink is applied to the screen with a spatula, the supply being replenished at intervals as required.

7.1.4 TYPES OF COMMERCIAL PRINTERS

A wide range of commercial printing machines is available, from hand operated laboratory types to fully automatic, large scale production equipments having a printing capability of some thousands of substrates per hour. For normal development and production work, rates of up to about one thousand printed parts per hour are obtainable employing semi automatic printers; in such machines, substrate loading and unloading are carried out manually, but advance and return of the substrate carriage and the squeegee printing cycle are automatic. The attainment of maximum printing rates in this case depends to a large extent upon the dexterity of the operator in loading and unloading the substrates.

In completely automatic machines on the other hand, automatic loading and unloading of the substrates enable the print cycles to be as rapid as one per second. Such high speed machines may incorporate a rotary substrate carriage which is automatically loaded from a magazine; one such machine is illustrated in Plate 7.2. In this type, the distance moved by the carriage between prints is considerably smaller than in the reciprocating type of substrate carriage.

By use of a rotary work table, cycle times as short as 1 second are made possible, so enabling rates up to about 3 500 prints per hour, or over 5 000 000 prints per machine per year, to be obtained. A single machine can thus produce about one million circuits requiring five separate prints per circuit (conductors, resistors etc.) per year.[1]

In other systems, a number of substrates are fed automatically from magazines to the print station, being then simultaneously printed through a multi-impression screen. In this type of system the problem of simultaneously registering a number of substrates has, of course, to be considered. One method of dealing with this aspect is to utilise the so-called 'snapstrate' substrates discussed in Chapter 3. A large area, acting as a single substrate, may here be printed in one operation through a multi-pattern screen, being then subdivided into single substrate units. In this way, the problem of registering a number of separate substrates beneath a multi-impression screen is eliminated; however, the problems resulting from greater variations in dimensional tolerances over the larger substrate area are increased.

Many printers, in particular the completely automatic types, incorporate a cut-out mechanism to arrest the machine should the substrate magazine become empty, or in the event of a misshapen substrate adhering to the underside of the screen. In the first case, ink is so prevented from being forced onto the underside of the screen, whilst in the second, damage to the screen itself is avoided.

7.2 THE SCREEN

The screen serves both to delineate the pattern to be printed, and to act as a metering device for the deposition of a precise quantity of the ink; the latter is of particular importance when printing resistors, where the resistance characteristics and their reproducibility are dependent on film thickness. The downward pressure of the squeegee during its traverse, forces ink into the mesh openings and deflects the screen into contact with the substrate. Some of the ink held in each of the mesh openings is then transferred to the substrate and remains there when the screen returns to its original position after the squeegee stroke. The discrete ink deposits so formed then coalesce to form a coherent film. The volume of ink held in each mesh opening depends on the mesh separation and diameter, and to some extent, on the thickness of the material used to form the printing configuration.

These characteristics, together with the manner in which they are retained during the life of the screen, and are reproduced from screen to screen, play a vital part in ensuring that precise and reproducible print characteristics are obtained.

Four types of screen are in use in thick film practice, namely direct emulsion, indirect emulsion, suspended metal, and solid metal; these are discussed in the following sections.

7.2.1 EMULSION SCREENS

In the formation of the direct type of screen a photographic emulsion is poured over the mesh, levelled with a doctor blade, and dried. The pattern is then defined by exposing the emulsion to an ultra-violet rich light source through a photographic mask to harden the areas to be retained on the screen. After exposure, the non hardened areas of the pattern are washed out by means of a developer solution, so defining the areas corresponding to the pattern to be printed. If a relatively small quantity of emulsion is applied this will merely

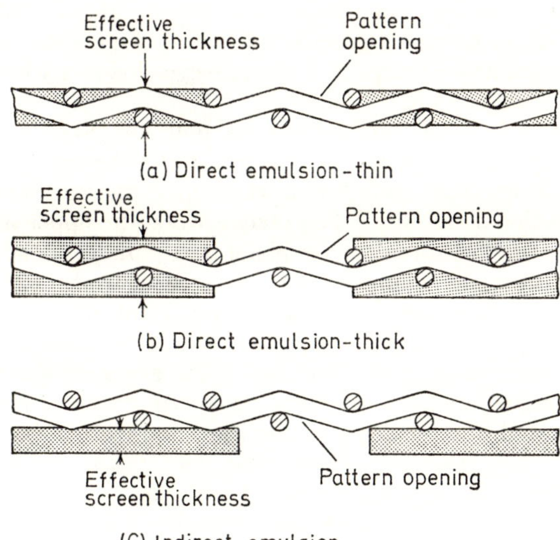

Fig. 7.2. Types of Emulsion Screen

block the mesh openings, but will not affect the screen thickness or the volume of ink delivered by each mesh opening (*see* Figure 7.2 (a)). Larger quantities will, however, increase both the effective screen thickness and the volume of ink delivered (*see* Figure 7.2 (b)).[2]

In the indirect emulsion type, a light sensitive film backed by a gelatine coated polyester sheet is used. The photographic exposure, development and washing processes are similar to those used for the direct emulsion type, but are carried out on the light-sensitive film/polyester sandwich. When developed, the wet film is transferred to a screen to which the gelatine adheres. After drying, the polyester backing is removed and the screen is ready for use. The indirect emulsion will increase the overall thickness of the screen and hence is a significant factor in determining the volume of ink delivered by the screen (*see* Figure 7.2 (c)).

In comparison with the direct emulsion screen, the indirect type gives greater uniformity of emulsion thickness and superior line definition, so permitting increased control over print characteristics to be obtained. The adhesion of the indirect emulsion screen to the mesh is, however, relatively poor compared with that of the direct type. Hence, the latter is less readily damaged and as a result, is capable of a longer operational life in normal usage.

The life of a direct screen is typically of the order of 20 000 prints, while that of an indirect screen may only be about 5 000 prints. The realisation of such figures is, of course, dependent on careful usage, particularly in respect of the squeegee loading which should be as light as possible.

The mesh materials commonly used, both for direct and indirect emulsion screens, are nylon and stainless steel. The mesh is mounted on a cast aluminium alloy frame which gives the required rigidity and dimensional stability. For precise work, the upper and lower surface are often ground to be plane and parallel to better than 0.002 in. Since the volume of ink deposited depends on the thread diameter and mesh opening area, it is important, if precise patterns are to be deposited, to employ meshes having as small a tolerance on these factors as possible. For the highest accuracy it is necessary to utilise mesh having a tolerance on aperture size of better than $\pm 3\%$. It is also important that the warp and weft of the mesh are uniformly perpendicular, and that the mesh is free from imperfections such as nicks or raised portions.

When the mesh is bonded to the screen framework it is necessary to ascertain that the mesh is not distorted, and that the perpendicularity of the warp and weft are retained. This will enable straight edges of circuits to be aligned uniformly along the length of the warp or weft over the full screen area, and will therefore minimise the effect known as 'saw-toothing'. This results when the edge of a straight emulsion pattern is laterally displaced from one mesh row to the next. Such an effect will, of course, militate against the production of high precision, fine-line prints.

Many users prefer to lay out circuit designs so that the pattern is generally parallel to the warp and weft of the mesh, the line edges, as defined by the emulsion pattern, coinciding with a wire so as to reduce saw-toothing to a minimum. However, some users claim that improved control over the volume of ink delivered per unit pattern area is obtained if the mesh is oriented at 45° to the pattern.[3]

In order to obtain reproducible printing characteristics it is important to control the screen tension closely, both during the life of the screen, and from screen to screen, since this factor governs, to a great extent, both squeegee pressure and breakaway distance. It is also important to ensure uniform tension over the entire screen surfaces, especially where large substrates or multiple images are being printed. In addition, close control of screen tension, will prolong the life of the screen, as well as reducing 'coining'. This is a permanent deformation of the screen over the edges of the substrate; this may occur by a combination of low screen tension and high squeegee pressure.

Screen tension is dependent both on the size of the screen frame and on the wire diameter. Thus, deflection for a given force will be greater in the case of a large frame and for a fine mesh. Since coarse meshes will deflect less than fine meshes, the latter will tend to have a shorter life. Moreover, fine meshes have comparatively small openings (0.002 in for a 325 mesh[4]) and therefore require a greater squeegee pressure to force the ink through the screen. As a result, fine meshes tend to have a relatively short life compared with coarse meshes and are therefore employed mainly when fine-line printing is required. In this context, a fine screen may be defined as one having a mesh size of about 325 or greater; such screens are employed, for instance, in printing fine-line conductors with a width of below 0.01 in. Standard conductor tracks are normally printed with 200 to 325, and resistors with 105 to 165, mesh screens.

Many thick film circuit producers prefer to purchase framed screens from specialist manufacturers, often with the printing pattern already defined. The control of such factors as initial screen tension and mesh framework alignment is therefore outside their control. It is, however, desirable to have an appreciation of these factors so as to be able to specify the type of screen for any particular application, and in order to obtain the maximum life from the screen in production useage.

Tension can be measured by means of a simple gauge, and reproducibility of tension from screen to screen, as well as gradual changes of tension with time, can therefore be readily checked. Screen tension gauges rely on the measurement of deflection of the screen under a specific applied force, the values being related to the screen area. In one type of commercial gauge, the force is applied to the centre of the screen employing a spherical contact tip of about 1 in radius. The deflection is measured by means of a 0.001 in dial gauge, the tip of which bears on the side of the screen opposite to the deflecting surface.

7.2.2 METAL MASKS

With the advent of flip chip and beam lead semiconductor devices (see Chapter 9), the necessity for improved techniques for the production of fine-line conductor patterns has increased. In order to be compatible with such devices the width of a conductor line must often be as small as 0.002–0.004 in, with a similar order of separation between the conductors.

Standard wire mesh emulsion screens limit the routine production of conductors to approximately 0.01 in width with 0.01 in spacing. Under closely controlled conditions, and utilising specially formulated fine-line conductor inks and fine mesh screens, the probable lower limit is of the order of 0.005 in conductor width and spacing. However, the development of metal masks used in conjunction with in-contact printing techniques has enabled thick film conductor patterns to be produced to the required dimensions.[5,6]

Two types of metal masks are available, namely suspended (or indirect) and solid (or direct).

In the suspended type, (illustrated in Figure 7.3) a metal foil of 0.001–0.002 in thickness is chemically etched to the required

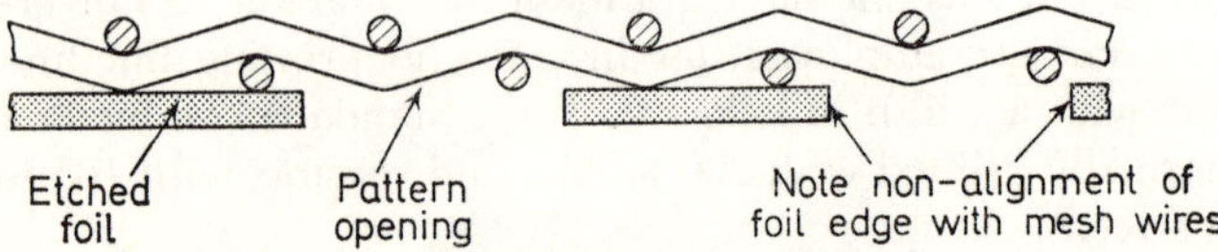

Fig. 7.3. Suspended Etched Metal Screen

pattern and is then bonded to a stainless steel mesh supported under tension on a screen frame. This technique is basically similar to that used in the production of indirect emulsion screens; the latter, it will be recalled, have a relatively limited life. In comparison with the longer lasting direct emulsion type however, the indirect metal mask enables a significant improvement in fine-line print precision to be obtained by virtue of the close control which can be exerted over foil thickness and wire diameter. Additionally, improved screen life is obtained since the metal structure is more resistant to abrasion and paste solvents than an emulsion pattern; however, as with emulsion screens, overstressing, with the danger of coining must be avoided.

Although the suspended metal screen yields a significant improvement in fine-line print capability, pattern tolerances attained are limited by the difficulty of aligning the etched image precisely with the wire mesh, since immediate bonding occurs on contact of the etched foil to the mesh during the fabrication process.

This problem is overcome in the solid metal etched mask. This comprises a bimetallic sheet in which the circuit pattern is etched

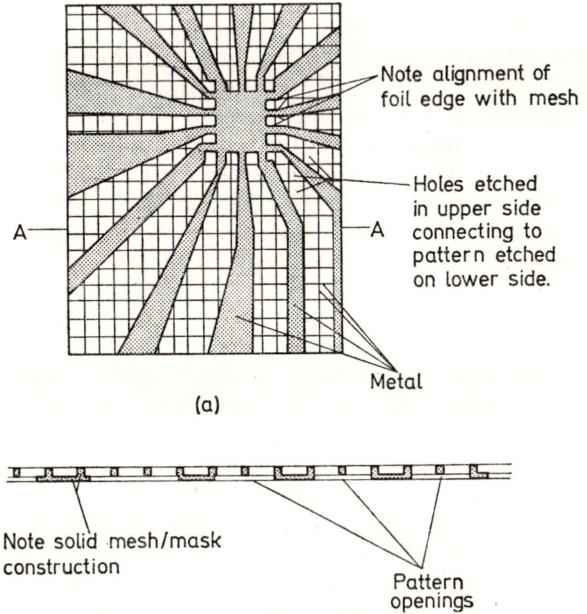

Fig. 7.4. Solid Metal Etched Mask

on the lower side and a grid of holes connecting to the pattern is etched on the upper side, as illustrated in Figure 7.4. Since the two patterns can be accurately registered under a microscope before etching, a significant improvement in line tolerance can be attained, thereby enabling finer lines to be printed. A further major advantage of the solid etched metal screen over all other types is its potentially longer life. This results since the pattern is supported by a solid mesh, and is thus more resistant to stressing than are other types, where the pattern is mounted on a woven wire mesh which is more readily distorted.

Since the metal mask is stiffer than a mesh screen and will not deflect as much, a smaller clearance between the mask and the substrate is required. In the case of the very fine lines necessary for beam lead and flip chip interconnections, the mask is, in fact, placed in contact with the substrate. This avoids the running together of closely spaced conductor prints which may occur even with very small mesh-substrate clearances. Contact printing necessitates the use of a printer having a facility for peeling apart the substrate and screen after the printing stroke.

It is claimed that, by the use of a three layer metal screen mask, lines as small as 0.001 in with clearances as low as 0.003 in and line width tolerances of ± 0.0002 in can be produced.[7] This type of mask comprises a sheet of metal such as copper, beryllium–copper, bronze or brass having on each side an electroformed layer of nickel. The multilayer structure is mounted on a precision machined frame under tension. The pattern configuration is electroformed in the lower layers, the upper one being provided with ink orifices which serve to meter precise quantities into the core and lower layers. The dimensions of the ink orifices and pattern cavities can be held within ± 0.0002 in, and the mask thickness can be closely controlled to enable various print thicknesses to be obtained.

In contrast with mesh screens, in which varying thicknesses of emulsion can cause differences in print thickness, the use of in-contact metal mask printing of resistor patterns is reported to confer increased reproducibility, owing to the more uniform print thickness yielded by the controlled thickness of the metal masks.[5] The improvement in reproducibility of characteristics is greatest, as would be expected, for small resistors. For a 0.020 in $\times$ 0.020 in resistor pattern, the variation obtained for metal screen printed resistors was 13%, compared with 26% for stainless-steel mesh printed resistors. In the case of resistors of 0.040 in $\times$ 0.040 in, variations for the metal mask printed specimens were 8.5–11%, as compared with 12–18% for those produced with wire mesh screens.

7.3 SQUEEGEE

It is important to consider the nature of the interface between the squeegee and the screen during the printing process, since this primarily governs the manner in which the ink is forced through the screen onto the substrate. The main factors to be considered are: the geometry of the squeegee blade, the physical properties of the squeegee, squeegee pressure and squeegee speed.

In order to obtain uniform transfer of ink through the screen, it is important that the printing edge of the squeegee should remain parallel to the screen throughout the printing stroke, (the screen should also, of course, be parallel to the substrate). Thus, in addition to ensuring that the squeegee is correctly adjusted relative to the screen, the squeegee blade should be checked for wear periodically. It should, in fact, be replaced at regular intervals before any appreciable wear occurs

In this connection, the squeegee blade should be of a material inert to the solvents commonly used in thick film inks. Neoprene is often used, although this material may swell in contact with some inks. Other materials used are polyurethane or PTFE, both of which are relatively inert and give long operational life. A medium hardness material is normally employed since a hard blade cannot distort sufficiently to conform to small irregularities in the screen, and therefore does not give a satisfactory wiping action. In contrast, a soft material, whilst giving a satisfactory wiping action, normally has a high wear rate and may, therefore, give non-uniform printing characteristics after a relatively short period. A sharp edged squeegee blade giving an attack angle of 45° is often recommended. Under certain conditions, however, a blunt edged, relatively soft blade has been shown to wear more slowly and print more consistently over longer periods than does a sharp edged blade.[8]

It is generally accepted that the squeegee pressure should be as low as possible, consistent with obtaining a printing stroke which just wipes the print pattern areas free of ink, and leaves a thin film on the blocked areas of the screen.

The printing force exerted depends on the screen tension and breakaway distance, and is normally in the range up to about 12 lb. Excess pressure may cause irregular print edges and, in the extreme case, may cause permanent deformation of the screen over the edges of the substrate.

The traverse speed of the squeegee is normally in the region of 1–10 in per second. A relatively low speed is generally preferred although at too low speeds, prolonged contact time between the screen and substrate can cause loss of edge definition of the pattern. At high printing speeds, entrapment of air in the ink may occur. This causes variations in the volume of ink delivered through each mesh opening, as well as possible disruption of the film during subsequent firing due to evolution of entrapped air. The interaction between these variables has been evaluated in order to devise a simple routine permitting conditions to be optimised for any specific printing medium.[8]

7.4 PASTE CHARACTERISTICS

The rheological characteristics of the paste, viz viscosity, surface tension and thixotropy, play, of course, an important role in determining the quality of the print. These factors affect the degree of wetting of both the substrate and the screen by the paste and hence determine, to a great extent, the proportion of the paste column in each mesh opening which is transferred to the substrate by the squeegee.

It is thus important to maintain these factors constant if reproducible results are to be obtained in practice. Pastes should be thoroughly stirred to ensure that the solids are completely dispersed, particularly after storage or when blending resistor compositions to obtain intermediate resistance values.

Solvent loss can be counteracted by addition of the recommended solvent. Care should be taken, however, to avoid over dilution, particularly of resistor compositions for which considerable variation in ultimate resistance characteristics would result. Viscosity can be checked by means of a rotating spindle viscometer, the paste being held at a standard temperature during measurement. To avoid variations in the rheological characteristics of the paste in practice, it is desirable that printing is carried out in a temperature controlled area.

7.5 FACTORS DETERMINING PRINT QUALITY

Having discussed the manner in which the component parts of the printer and the physical properties of the paste can affect the print

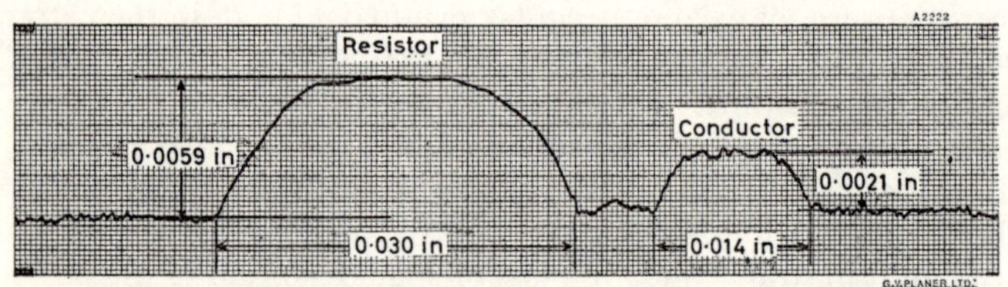

Fig. 7.5. Thickness Profile Trace of Resistor Track. (*G. V. Planer*)

characteristics, we are now in a position to summarise their effects on the quality of the ultimate print.

The main characteristics of the print to be considered are film thickness and pattern definition. Fired film thickness should normally be in range of 0.0003–0.001 in, depending on the particular ink used. The attainment of uniform and reproducible print thickness is, of course, of prime importance in the case of resistors since this is a factor which controls track resistance. Methods and equipment for the measurement of thickness are discussed in Chapter 3; Figure 7.5 shows the thickness profile traces of a resistor and conductor track obtained employing one such instrument.

7.5.1 FILM THICKNESS

Thin prints result if

the viscosity of the paste is too low,
the squeegee pressure is too high,
the stroke speed is too high,
the breakaway distance is too small,
the angle of attack is too small, or
the squeegee blade is too hard.

The converse of each of these parameters will, of course, result in the production of relatively thick prints.

Non uniform deposition may also occur, with a variation in print thickness from end to end, or from side to side of the track. End to end non uniformity may result if

the speed of the squeegee stroke is not constant,
the substrate and screen are not parallel,
there is insufficient paste on the screen, or
the breakaway distance is too small.

Side to side non uniformity may result if

the substrate and screen are not parallel,
there is insufficient paste on the screen,
the squeegee pressure is too low, or
the squeegee blade is worn.

Non uniform deposition may also occur if the viscosity of the paste is too high. While care may have been taken in adjusting the overall screen to substrate parallelism, the latter may, of course, be affected by local variations in, for example, the camber of the substrate, with the result that localised non uniformity of print thickness may occur.

7.5.2 PATTERN DEFINITION

A poorly defined print may be characterised by ragged edges, an incompletely filled pattern or a smeared image.

Ragged print edges may be caused by

a badly made or worn screen,
a worn squeegee blade,
insufficient squeegee pressure, or
a rough substrate.

An incompletely filled pattern may result if

the screen is badly made or some of the meshes are blocked,
the viscosity of the ink is too high,
the squeegee pressure is too low,
the squeegee blade is too hard or is worn,
the angle of attack is too low, or
the print speed is excessive.

A smeared image may be caused if

the viscosity of the ink is too low,
the breakaway distance is too small
the print speed is too slow,
the squeegee blade is worn,
the squeegee pressure is excessive,
the screen is dirty, or
the screen tension is too low.

A damaged or excessively worn screen results if

the squeegee pressure is excessively high,
the breakaway distance is too great,
the emulsion is not fully cured, or
the emulsion is soluble in the solvents in the ink.

REFERENCES

[1] Hughes, D. C., 'The Screen Process Printing Machine'. *Electronic Packaging and Production*, 67–73, (Feb., 1968).

[2] Reimer, D. E., 'The Direct Emulsion Screen as a Tool for High Resolution Thick Film Printing'. *Proceedings, IEEE 21st Electronic Components Conference*, 463–467, (1971).

[3] Finch, R. G., 'Printing of Thick Film Microcircuits'. *Microelectronics*, 14–16, (Jan., 1968).

[4] BS 410, 1969. Test Sieves, British Standards Institution.

[5] Littell, W. C. Jr., 'Use of Metal Masks to Improve the Printing of Thick Film Geometries', *Proceedings of the International Society for Hybrid Microelectronics*, 227–232, (July, 1968).

[6] Hughes, D. C. Jr., 'Contact Printing Techniques in Screen Printing', *Ceramic Age*, 32–35, (Nov., 1969).

[7] Guildemond, J., 'Metal Screen Mask Advances Precision in Fine Line Printing', *Solid State Technology*, 54, (July, 1969).

[8] Savage, J., 'Factors Affecting the Quality of Screen Printed Conductors', *Thin Solid Films*, No. 4, 137–148, (1969).

8

Firing Procedures

Subsequent to the printing process, the units have to undergo the stages of coalescence of the newly printed pattern and drying to remove solvents before they are ready for the firing process.

8.1 INK COALESCENCE

Immediately after printing a pattern through a mesh type screen, the print is made up of a series of discrete ink spots each corresponding to a mesh opening in the screen. The substrates must therefore be allowed to stand at ambient temperature for some five minutes to permit the ink to flow sufficiently to form a coherent, level film in which the mesh pattern is no longer visible. The time required for this coalescence depends on the nature of the composition, being longest for highly thixotropic inks, whilst in other cases, adequate flowing may occur within a few seconds of printing. In the case of prints produced by meshless metal masks (see Chapter 7), no such time for coalescence need be allowed.

8.2 DRYING

The second stage of the process is that of drying the printed film to remove the more volatile ink components. Temperatures of the order of 70–150 °C are commonly employed, for periods ranging from 15 to 30 minutes. The conditions depend upon the composition as well as on the method of drying adopted.

Drying is usually carried out using infra-red sources, positioned both above and below the substrate to provide even heating. This is most conveniently carried out in a tunnel oven through which the substrates travel on a moving belt. In a production facility, the drying tunnel often links the printer unloading station with the furnace loading station to form a continuously moving line. If the ink used requires an appreciable flowing time, this must, of course, be allowed for in the advance rate to the drying unit.

The use for this purpose of infra-red energy with a wave length of 3 microns has been advocated;[1] this results in adequate penetration within the ink, promoting bulk heating and preventing entrapment of solvents by the formation of a surface skin of dried material. As already indicated, uniform heating is also promoted by heating the substrate from below.

Alternatively, the printed elements are sometimes dried by advancing them through an oven with forced hot air circulation. In this method there may, however, be some risk of solvent being trapped beneath the outer skin so formed. In addition, airborne dust particles may be more readily picked up by such a drying film, unless the air in the oven is first passed through a filter.

8.3 FIRING

After drying, the circuits are fired to develop the ultimate conductive, resistive or dielectric characteristics of the compositions. The firing process involves two stages, taking place consecutively in the same furnace. During the first stage the remaining solvents are volatilised and the organic binder is carbonised and oxidised. These processes occur at temperatures of up to about 450 °C and require an oxidising atmosphere.

During the second stage, in which the circuits are taken to the maximum firing temperature, which may be up to 1 000 °C, the glass component of the ink melts to form a vitreous medium which serves to consolidate the printed layer, and also promotes adhesion to the ceramic substrate. In this way the final characteristics of the ink are developed. A chemical reaction may occur as in the case of palladium–silver resistor compositions (see Chapter 5), or the resistive, conductive or dielectric material may merely become dispersed within the molten vitreous phase. As the temperature becomes lower during passage through the furnace, the molten glass solidifies to form a coherent film. Time-temperature firing curves are illustrated in Figure 8.1 for typical resistor, conductor and encapsulant glazes. The dwell time at peak temperature may vary from one minute to more than 15 minutes, both this and the peak temperature being dictated by the type of composition being fired.

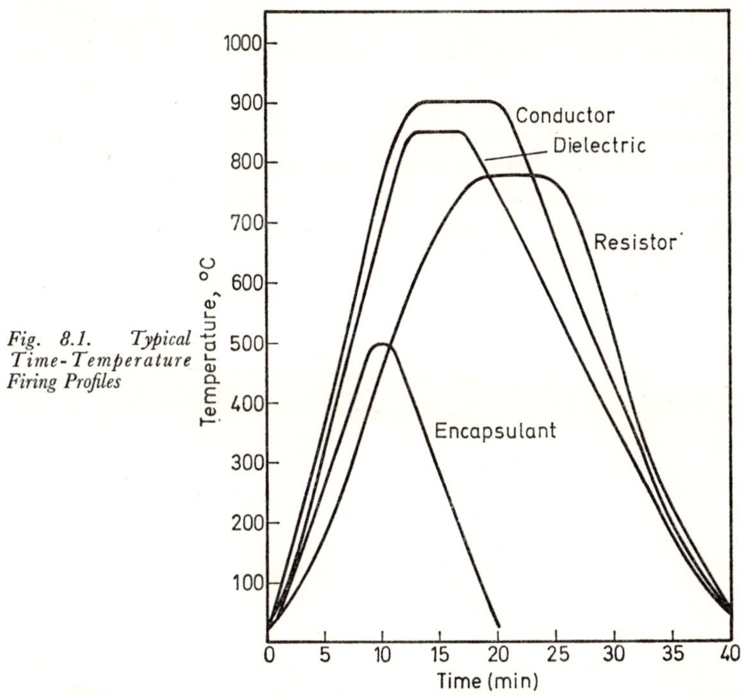

Fig. 8.1. Typical Time-Temperature Firing Profiles

8.4 FURNACES

Thick film circuits are, in nearly all cases, fired in a multi-zone, continuous belt furnace in which the speed of the belt and the temperature in each zone are controlled to correspond to the time-temperature curve appropriate to the material being fired.

A wide range of commercial furnaces is available, many of which conform to a similar basic design. Typically they comprise a rectangular or D-shaped muffle tube running the length of the surface, and through which passes a continuous wire mesh belt. The muffle is heated externally on its four sides by resistance windings supported on grooved ceramic tiles. The heating elements are generally wound from Nichrome (nickel–chromium) or Kanthal (iron–chromium–aluminium) wire. A furnace of this general type is shown in Figure 8.2.

Fused quartz muffles are often used in such furnaces, since this material is inert to oxidising atmospheres up to 1 000 °C. It is more than 80% transparent to infra-red radiation in the temperature range of 600–1 100 °C. The small degree of opacity serves to diffuse the radiation incident on the muffle from individual heater windings, and so improves the temperature uniformity within the muffle.

Metal muffles made of high nickel content alloys have also been used, particularly in cases where the walls of the muffle have to be pierced to provide individual vents for each section of the furnace.[2] Such alloy muffles oxidise to some extent at temperatures in excess of 600 °C and, although the oxidation products are well adherent to the metal, there may be some risk in prolonged use of reaction products becoming detached and contaminating the circuit layers being fired.

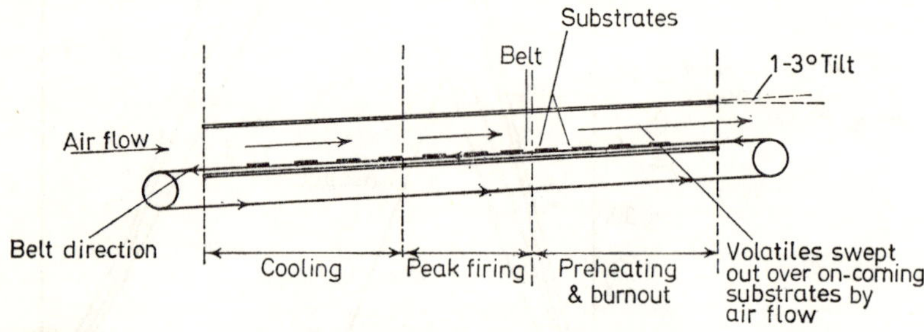

Fig. 8.2. Air Flow Through Inclined Furnace

Laboratory and pilot plant furnaces generally have three or four separately controlled furnace zones with two to four inch wide belts. On the other hand, large production equipments may have as many as twenty-four separately controlled zones, with belts as wide as sixteen inches. Some furnaces are provided with water cooling facilities at the exit end, to permit more rapid cooling of the plates in the furnace.

Each furnace zone is normally heated independently to enable the required temperature profile to be established. In most commercial furnaces the temperature of each zone is independently sensed and controlled by means of a separate thermocouple and control system. The latter may comprise proportional band controllers with automatic reset, or proportional controllers with integral and derivative action. These modulate the output of silicon controlled rectifiers in the heater circuit in response to the temperature controlled output. In the case of furnaces having wide belts, say of about 8 in width or more, separate systems for controlling the temperature across the belt may also be provided.

The positioning of the thermocouple tips is critical for accurate temperature control, particularly since it is important that the furnace characteristics are maintained constant in both the case of a lightly and a fully loaded belt. Such conditions can exist in production if there is a temporary reduction in the rate of supply of substrates to the furnace, or if a small number of test samples are followed by a full batch of production plates. One class of furnace incorporates a thermosensitive hearth system[3] designed to maintain the same temperature profile, regardless of the belt loading. In this case the thermocouple is located so as to sense the temperature of both the top and bottom heaters, as well as that of the substrates, being capable therefore of responding instantly to the presence of the latter.

Utilising such control systems, temperature variations in commercial furnaces are typically held to ± 1 °C across the belt, and from ± 1–2 °C along the belt at temperatures of the order of 1 000 °C.

The belt generally consists of a wire mesh of a high nickel content alloy such as Inconel, selected for its relatively good corrosion resistance properties under the oxidising conditions encountered in the furnace. Whilst some oxidation will occur in normal useage, the fact that the belt is underneath the substrates mitigates to some extent the danger of contamination of the plates by the corrosion products. Nevertheless, many furnaces incorporate some means of cleaning the belt, generally using a rotary brush or vacuum cleaner applied to the belt on the return side of the furnace.

A fine weave mesh is generally used so as to make the belt thermally massive relative to the substrates, and to permit small substrates to be placed directly on it. Belt speed control is important if precise temperature-time profiles are to be maintained in prolonged use. Speeds are usually controllable within the range of 0.5–12 in per minute, an accuracy of $\frac{1}{4}$% of the speed range being obtainable under varying conditions of useage in some commercial furnaces.

Control of the atmosphere and its rate of flow through the furnace is important, particularly in the case of palladium–silver resistors in which a critical oxidation-reduction reaction occurs during firing.

A simple method of maintaining a reasonable degree of control in the flow rate is to incline the furnace at an angle of between 1 and 3 degrees, so inducing a flow of air through the furnace towards the loading end, as illustrated in Figure 8.3. This ensures that the circuits being fired in the hottest zone always encounter an uncontaminated supply of air, since all combustion products will be swept out of the furnace over the oncoming circuits at a relatively low temperature. For this purpose many thick film furnaces incorporate a variable angle tilting mechanism. This method provides a somewhat limited degree of control over the furnace atmosphere, especially since it relies on the use of ambient atmosphere.

It is important to ensure that laminar flow of the gas occurs through the furnace, particularly in the region adjacent to the moving belt, since the development of uncontrolled turbulence in this area can lead to appreciable temperature fluctuations, and hence to poor reproducibility of fired film characteristics.

For greater atmospheric control the ends of the furnace tube can be closed and gases of predetermined composition, particularly in respect of water vapour content, may be fed in at a controlled pressure and rate of flow. Again it is important to ensure that this method does not lead to the development of uncontrollable turbulence within the furnace tube.

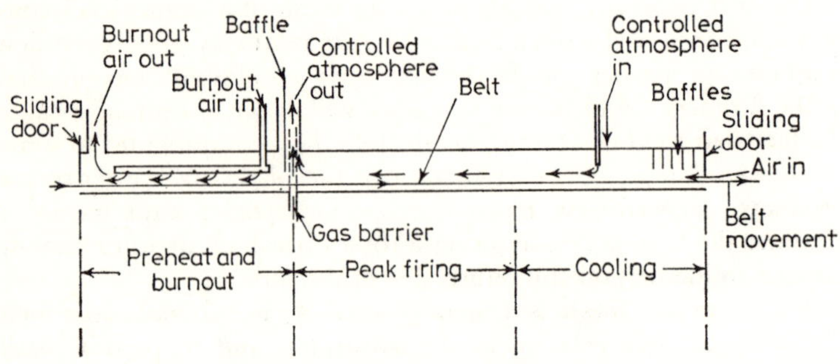

Fig. 8.3. Controlled Atmosphere Furnace

Some furnaces have facilities for maintaining different atmospheres in the preheat zone in which oxidation of the organic materials occurs, and in the sintering zone in which the final firing of the composition occurs, as illustrated in Figure 8.3. To accomplish this, protective curtains, normally of nitrogen, are arranged at each end of the relevant furnace zones and again, laminar flow of gas, particularly in the sintering zone, is provided. Such protective barriers should be at a negative pressure with respect to the furnace atmospheres, to ensure that the gas flow through each section is removed at the exit end through the barrier, and does not pass into the next section.

In some furnaces, the temperature control system includes temperature monitors at points both within and outside the atmosphere flow. This is to permit changes in the cooling effect due to variations in the laminar flow to be detected and corrected.

8.5. DETERMINATION OF TEMPERATURE PROFILE

The temperature–time profile of the furnace under working conditions can be determined by attaching a long thermocouple to the moving belt and recording the indicated temperature as it passes through the muffle. In order to evaluate the variations between no load and load conditions, the thermocouple tip is, in the latter case, placed in contact with one of the circuit substrates, the belt being loaded with the number of substrates to be processed in practice. In the case of a wide belt, multiple thermocouples are used to measure any temperature variation across the belt, and for example, three may be used in the case of an 8 in belt.

Fine wire thermocouples are employed for these purposes to avoid any undue effect on the thermal mass of the belt. They are usually of the platinum–platinum/rhodium type, since this does not suffer from cold working effects due to the handling and bending of the wires during repeated profile determinations. Cold working of thermocouple wires would entail the risk of inhomogenous sections being formed within the wires and spurious EMFs could hence be developed at such points during the passage through the furnace, with consequent errors in the indicated temperatures.

REFERENCES

[1] Beck, J. H., 'Firing Thick Film Integrated Circuits', Technical Publication, BTU Engineering Corpn.

[2] Brennan, H. F., 'Furnace Equipment for Thick Film Production', *Electronic Equipment News*, 58–63, (May, 1969).

[3] 'Getting Results in Thick Film', BTU Engineering Corpn.

9

Hybrid Circuits - Attachment of Chip and Other Discrete Devices

Previous chapters described techniques for the fabrication of circuits comprising printed conductors, resistors, and capacitors. The information in this chapter covers the types of discrete devices, both active and passive, which are available for incorporation in hybrid thick film circuits, together with the various methods of mounting and interconnecting such devices in the circuit pattern.

Semiconductor devices for use in hybrid thick film circuits fall into the two main classes of non packaged and packaged components. The first type includes standard chips, flip chips and beam lead chips, whilst the latter group comprises miniature plastic encapsulated devices, flat pack ICs, dual in-line packages (DILs), T05-type packaged devices and ceramic mounted leadless inverted devices (LIDs).

Discrete capacitors in the form of miniature ceramic or sintered tantalum chips are used in many applications. These types largely overcome the inherent limitations of screened and fired capacitors, namely their low capacitance/area characteristics and the additional process steps necessary for their fabrication (see Chapter 6).

Other types of components such as resistor chips and miniature inductors can also be incorporated in hybrid thick film circuits.

The techniques available for bonding the chosen device to the circuit conductor pattern depend primarily on the metallurgical nature both of the device contacts, and of the thick film conductor material (see Chapter 4). Methods in widespread use include: re-flow soldering, thermocompression bonding, ultrasonic welding, alloy-formation bonding, and adhesive bonding.

In the following sections, the various types of devices and the methods employed for bonding them to the thick film circuit will be discussed in detail.

9.1 NON PACKAGED SEMICONDUCTOR DEVICES

These can be incorporated in the circuit in the form of back bonded chips, face bonded flip chips and beam lead chips.

In the back bonding, or 'chip-and-wire,' technique, as the terms imply, the rear face of the chip is bonded to the thick film substrate and individual wire connections are made successively between each contact area on the face of the chip and appropriate conductor areas on the substrates (see Figure 9.1(a)). In flip chip bonding, the chip is inverted so that its active side faces the substrate and the contact areas are then bonded simultaneously to corresponding areas on the substrate, (see Figures 9.1 (b) and (c)). For this purpose, raised pillars

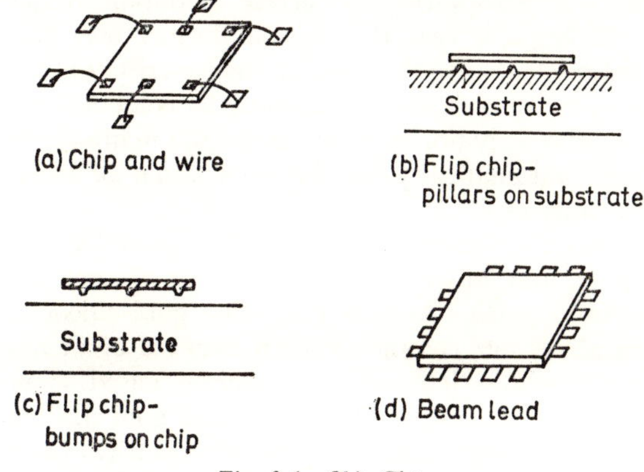

Fig. 9.1. Chip Types

or bumps are provided, either on the chip or on the substrate. The back bonding, and the latter type of flip chip bonding techniques may utilise virtually any type of standard transistor or IC chip and these procedures are therefore particularly versatile. For beam lead bonding the chip is produced with integral tape leads extending from each contact area beyond its periphery, and bonding to the substrate is carried out in a single operation (see Figure 9.1 (d)).

9.1.1 BACK-BONDED CHIPS

Virtually any type of semiconductor chip may be back-bonded to a thick film substrate employing eutectic alloy or adhesive bonding techniques. Leads can then be attached to the chip contact and thick film conductor areas by thermocompression or ultrasonic bonding. Circuits produced in this fashion must subsequently be hermetically sealed to prevent degradation of the semiconductor device characteristics in useage. Some types of chips are provided with a silicon oxide glassy passivation layer which provides a limited degree of environmental protection.

The eutectic chip bonding technique relies on the fact that the silicon–gold eutectic alloy melts at a temperature as low as 370 °C, compared with the normal melting points for gold and silicon of 1063° and 1404 °C, respectively. In this method, the thick film substrate comprising gold conductor pads is heated to about 400 °C. The silicon chip is then placed with its rear face in contact with the heated pad, and is 'scrubbed,' (i.e. lightly vibrated), over the surface. This motion results in the formation of the silicon–gold eutectic alloy which solidifies on subsequent cooling to form an ohmic contact having good mechanical bond strength. The action also breaks up and displaces any surface oxidation at the bonding interface and helps to seat the chip uniformly over the bonding area. In order to enhance the wetting characteristics of the bonding surfaces, the rear surface of the chip may be coated with a layer of gold. The scrubbing action may be carried out manually but is more often effected automatically by the application of mechanical or ultrasonic vibrations.

Bonding may also be carried out employing preforms of eutectic alloys of gold and other metals, including silicon, germanium, indium, gallium, antimony and tin. The gold–silicon and gold–germanium alloys are commonly used because of their respective compatibilities with silicon and germanium based semiconductor materials. Gold–gallium, gold–indium and gold–antimony alloys are normally employed to develop specific electrical characteristics such as the creation or prevention of additional p and n type junctions for particular applications. Gold–gallium and gold–indium alloys are used with p type materials, and the gold–antimony alloy with n type materials. Gold–tin alloys have particularly low melting points and are commonly used for applications in which a highly conductive joint between the dice and substrate must be formed at low temperatures. The melting points of the gold eutectic alloys, together with the corresponding weight percentage of gold in the alloys are listed in Table 9.1.[1]

Semiconductor dice can alternatively be back-bonded to the thick film substrate with either an electrically conductive or non conductive thixotropic epoxy resin paste. The former type is filled with gold or silver powder, whilst the latter may contain aluminium or asbestos to increase the thermal conductivity and lower the coefficient of thermal expansion of the adhesive. A metered amount of the resin is dispensed on to the substrate surface, often using controlled pulses of compressed air to force the resin through a nozzle. The chip is pressed into position on the surface of the resin which is then cured, either at room, or at elevated temperature.

The heat dissipation capability of chip devices may be increased by back-bonding them to a larger metal base, such as a gold plated Kovar tab, which is then soldered to the substrate pattern.

Table 9.1

ALLOYS UTILIZED FOR BACK BONDING

Eutectic Alloy	*Melting Point °C*	*Weight % Gold*
Si–Au	370	94
Ge–Au	356	88
In–Au	451	73
Ga–Au	341	84
Sb–Au	360	76
Sn–Au	232	10
Sn–Au	280	80

Connections between the contact areas on the chip and the substrate are made by bonding gold or aluminium wire leads, normally of about 0.001 in diameter, either by thermocompression or ultrasonic bonding.

In thermocompression bonding, the connector is positioned on the contact pad and heat and pressure are simultaneously applied by a bonding tool to cause plastic flow and molecular bonding of the two surfaces.[2] The temperatures used are well below the melting points of the joint components. For example, in the case of fine gold wires (0.001 in diameter) the bonding temperature is normally about 300–400 °C, the applied force of the order of tens of grams, and the bonding time about 1 second.[3] The temperature may be raised by heating the bonding head, or the substrate platform, or both.

Several variants of the thermocompression bonding technique are used in practice; these involve the use of different bonding heads, namely wedge, ball, stitch and eyelet bonding, as illustrated in Figure 9.2 (a)–(d), respectively.

In wedge (or chisel) bonding, the tip consisting of sapphire or tungsten carbide is in the form of an accurately radiused anvil, the radius normally being made two to three times the diameter of the wire to be bonded.

In ball bonding (also referred to as nail head bonding), the tip of the wire is passed through a vertical glass capillary tube and is melted by a hydrogen flame to form a ball. The latter is pressed onto the contact land by the end of the capillary tube to form the bond. After withdrawal of the capillary, the wire is cut with the flame, simultaneously forming a ball for the next bond.

Stitch bonding is a combination of wedge bonding and ball bonding. The wire is again fed through a capillary tube and, on issuing from the latter, is bent through a right angle beneath the tip, the horizontal portion taking the place of the flame formed ball.

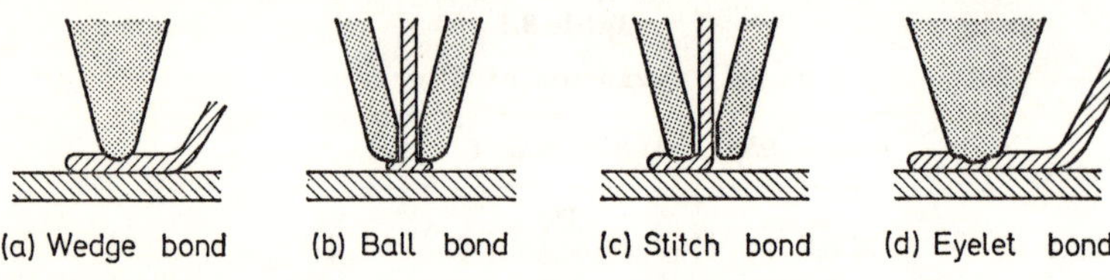

Fig. 9.2. Thermocompression Bonding Heads

After bonding, the capillary is withdrawn and moved to the next contact area to repeat the operation. This sequence is continued several times until the wire is finally cut.

Wedge and stitch bonding can be carried out using gold or aluminium wire, while ball bonding is limited to gold only, since a ball cannot successfully be formed in this way from aluminium. By virtue of their shape, however, ball bonds generally have a higher strength than wedge or stitch bonds, which reduce the cross sectional area of the wire and hence weaken it.

The so called eyelet bonding method, a variant of conventional thermocompression bonding, overcomes these difficulties and allows the formation of reliable bonds between aluminium wire and aluminium pads.[4] In this method, which derives its name from the eyelet shape of the resulting bond, the bonding head is specially designed to induce lateral flow of the aluminium to occur during bonding. This results in break down of the surface oxide layer, permitting intimate molecular bonding between the two components. The method is of special use where aluminium wire is to be bonded to aluminium pads, and in particular, where device reliability is of prime importance.

Ultrasonic welding depends on the production of ultrasonic shear energy at the interface between the two components to be united, the process here being carried out at room temperature. The energy is transmitted from a magnetostrictive transducer vibrating at a frequency of about 60 kHz by a short vertical welding tip. This presses against, and clamps together, the components to be joined.

The bond so formed is comparable to the seizure of two non-lubricated parts due to friction. Ultrasonic bonding is particularly effective in the case of aluminium to aluminium bonding since by this means, the oxide surface layers normally present are readily broken down.

Although gold wires are often used for connection to the aluminium pads on the semiconductor chip, their use is best avoided if the final assembly is to operate continuously at temperatures in excess of 150 °C. Under such conditions, various gold–aluminium intermetallic compounds, including a purplish–black material (the so called 'purple–plague' phenomenon) will form at the gold–aluminium interface. The compounds are generally brittle, may have a low electrical conductivity, and may cause degradation of the contact to the point of ultimate failure. The rate of reaction at room temperature is extremely low, but becomes appreciable at temperatures above 200 °C. In applications where the device is to operate at temperatures of this order, it is therefore preferable to use aluminium wire connectors.

Although the use of an aluminium wire connected to a gold-containing thick film composition may appear to transfer the problem from the semiconductor chip to another position, practical evidence points to the presence of silicon as speeding up the reaction.[5] It is generally considered to be quite acceptable to use aluminium wires connected to, for example, gold–plated header terminal pins or thick film gold conductors.

In order to protect the aluminium contact pads from excessive oxidation during high temperature processes associated with die and thermocompression wire bonding, these processes must be carried out in a shroud of an inert atmosphere, normally of oxygen free nitrogen.

The major advantages of utilising conventional chips in hybrid circuit applications are their ready availability and low cost. There are, however, a number of important disadvantages to be considered. The major one is undoubtedly that all the front surface contacts, both on the chip and on the thick film substrate, have to be individually bonded. This process requires skilled operators and expensive equipment to produce an acceptable yield, and the bonding costs may go a long way towards counteracting the low cost of the chip itself.

The performance of the chip, as obtained from the manufacturer, is predictable, based as it is on normal parameter spreads. However, the characteristics of the device may be significantly modified during hybrid circuit assembly, both by the back-bonding and interconnecting processes. Such changes may be particularly marked if many chips are to be attached to the substrate by high temperature back-bonding, or by those thermocompression wire bonding techniques which necessitate holding the chips at elevated temperatures for prolonged periods. After chip attachment and interconnections are completed, each device must therefore be tested for satisfactory functioning. Some variation in device performance may occur moreover during subsequent packaging or encapsulation, if these involve the use of elevated temperatures.

The application of heat during back-bonding and thermocompression lead bonding may also cause irreversible changes in the values of the printed passive components on the thick film substrate. This may not, however, be significant if the circuit design permits final adjustment of the assembly to a functional value by subsequent trimming of resistors.

9.1.2 FLIP CHIP DEVICES

In flip chip bonding the semiconductor device is inverted and bonded in the face down position to the substrate interconnection pattern. Bonding is normally carried out by ultrasonic, thermocompression or reflow solder techniques, the bonds being made virtually simultaneously. In this manner the disadvantages of the large number of individually produced bonds required in the case of chip and wire bonding are obviated. In contrast to the latter method, however, the bonds cannot be individually tested by non-destructive methods.

In machines designed for flip chip bonding, the vertical welding stub generally comprises a fine axial capillary tube connected to a vacuum pump, so enabling the chip to be picked up and positioned on the substrate by means of the welding tip itself. A split-image optical system is employed to obtain a magnified view of the active underside of the flip chip superimposed on that of the substrate. Movement of the weld tip holding the chip then enables the two images to be brought into registration, so permitting the chip to be accurately located on the substrate prior to bonding.

The contacts may take the form of raised bumps on the contact areas of the chip (as illustrated in Plate 9.1), or of raised pillars on the conductor pattern. In the latter case, use can be made of virtually any type of standard chip (as in the case of back-bonding), and it is then necessary to match the pillar configuration to that of the contact areas on the chip.

Several methods of providing contact pillars on the substrate have been used. In one method, the pedestals are formed at the required locations by the printing and firing of pads of a platinum–silver

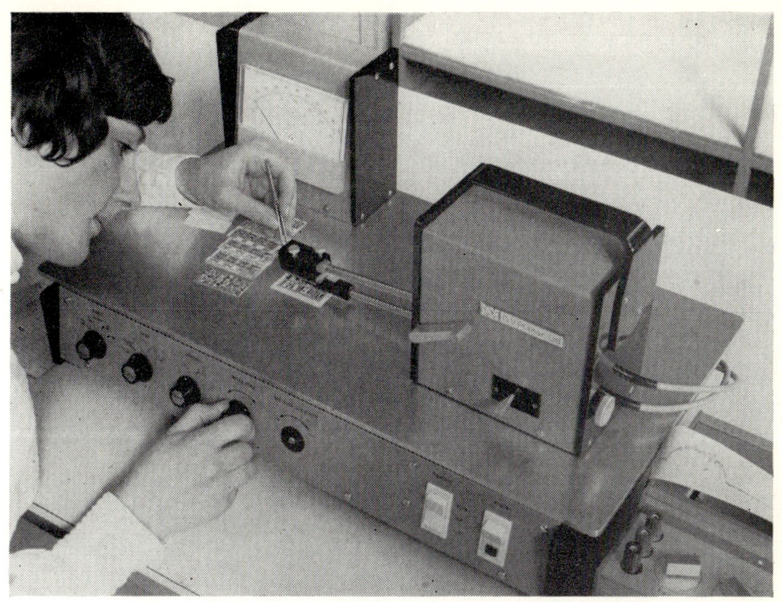

Plate 3.1. Surface Finish Test Instrument. (G. V. Planer)

Plate 3.2. Thick Film Substrates (Coors Porcelain)

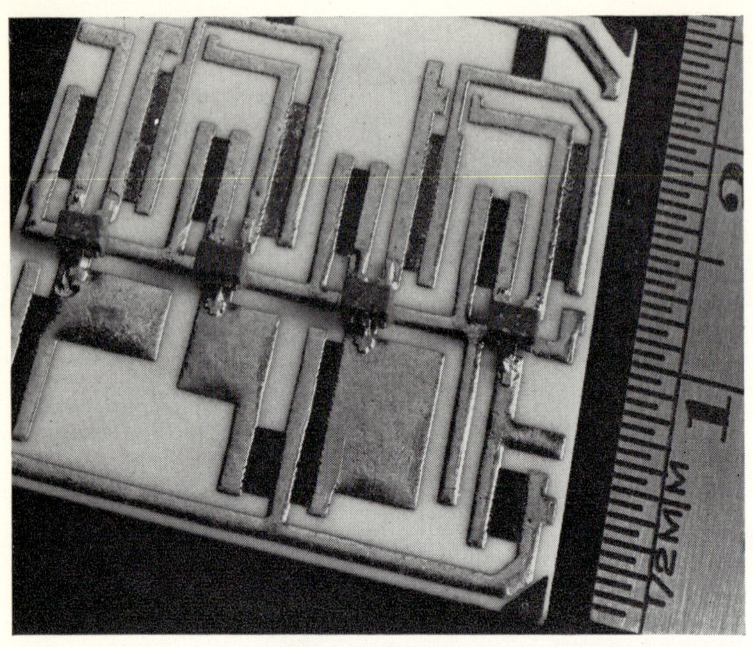

Plate 4.1. (above) Circuit with Tinned Interconnections (Ferranti)

Plate 7.1. Printer Illustrating Screen and Squeegee System (DEK)

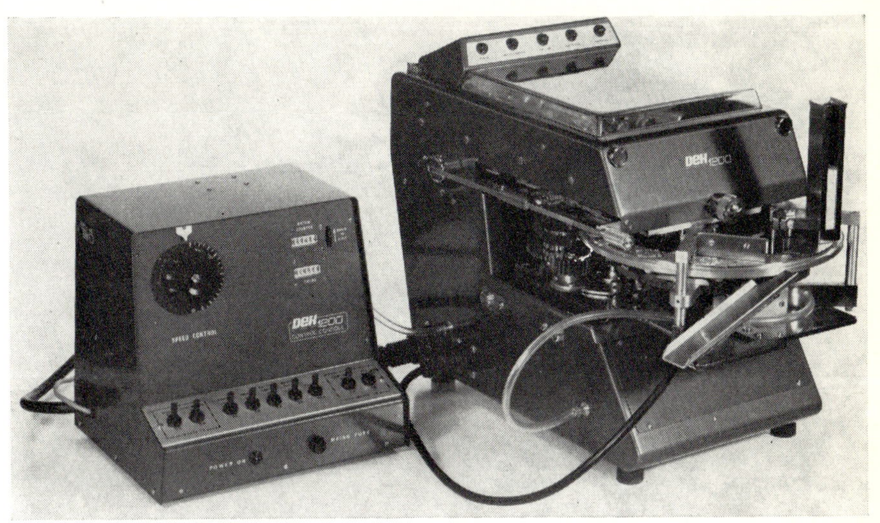

Plate 7.2. Automatic Printer with Rotary Substrate Carriage (DEK)

Plate 8.1. Thick Film Firing Furnace (Efco Royce)

Plate 9.1. Flip Chip Transistor (Welwyn Electrical)

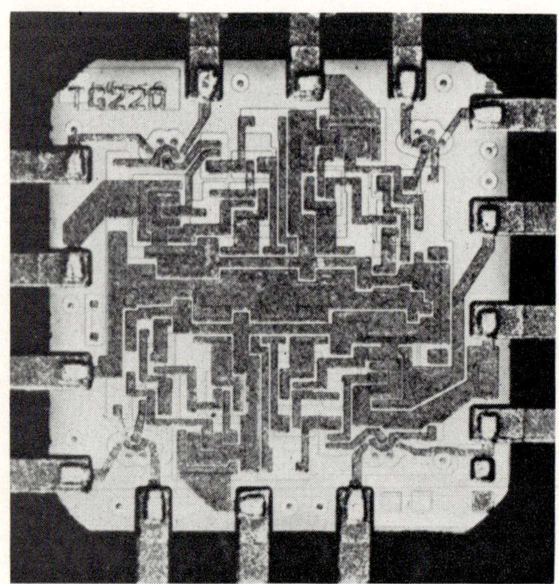

Plate. 9.2. 14-Contact Beam Lead IC (Raytheon)

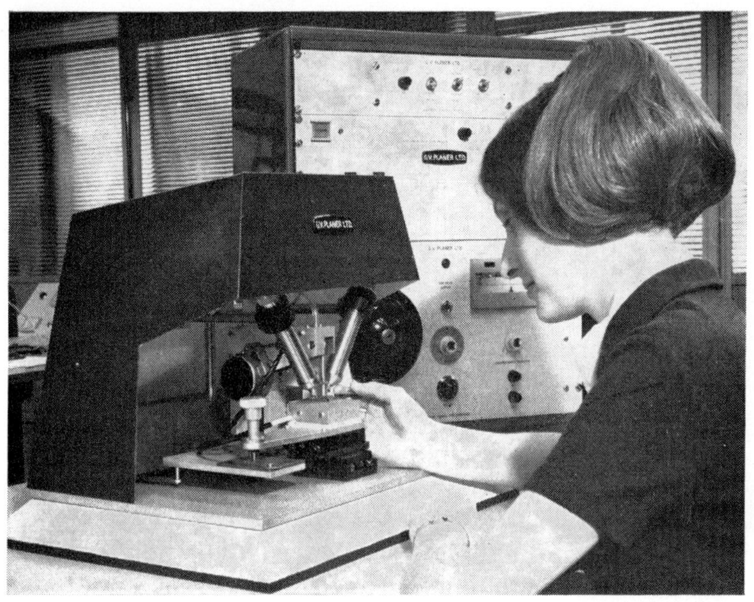

Plate 9.3. Semi-Automatic Hot-Gas Bonder (G. V. Planer)

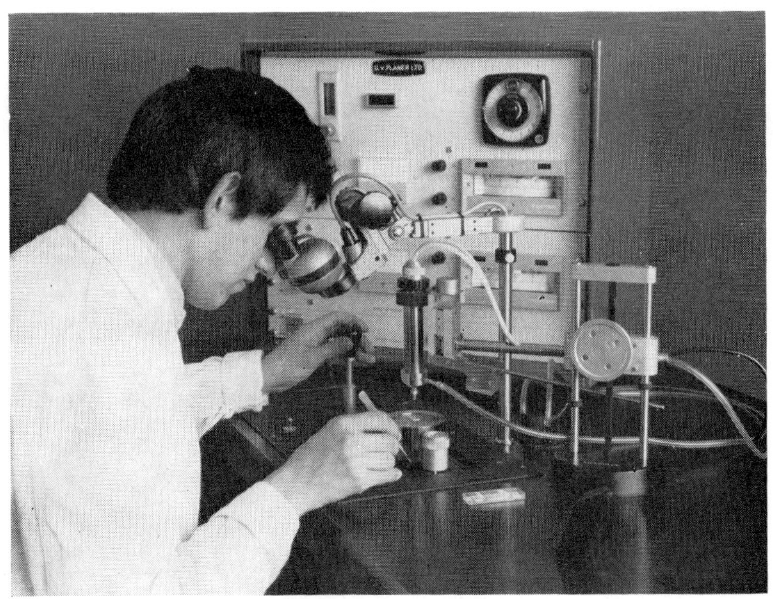

Plate 9.4. Manually-Operated Hot-Gas Bonder (G. V. Planer)

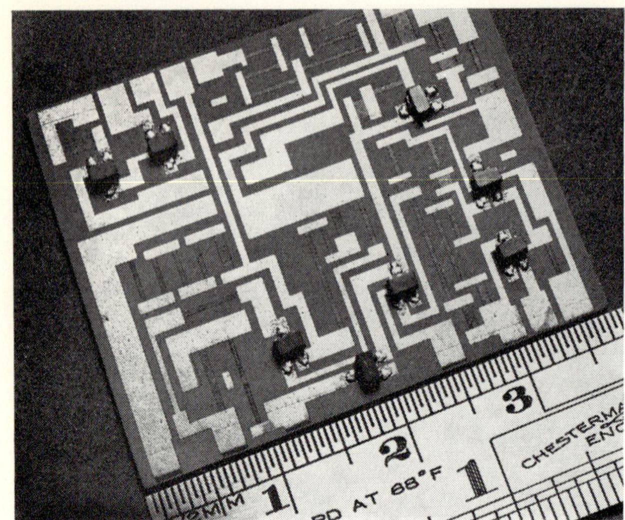

Plate 9.5. Plastic Encapsulated Transistors (Ferranti)

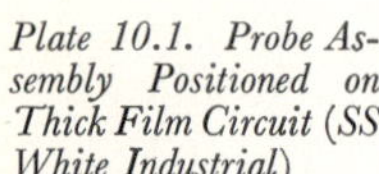

Plate 10.1. Probe Assembly Positioned on Thick Film Circuit (SS White Industrial)

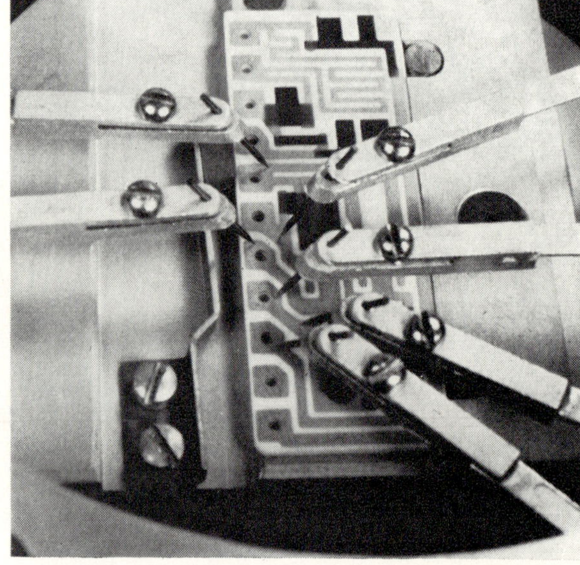

Plate 10.2. Four-Head Rotary Table Air-Abrasive Trimmer (SS White Industrial)

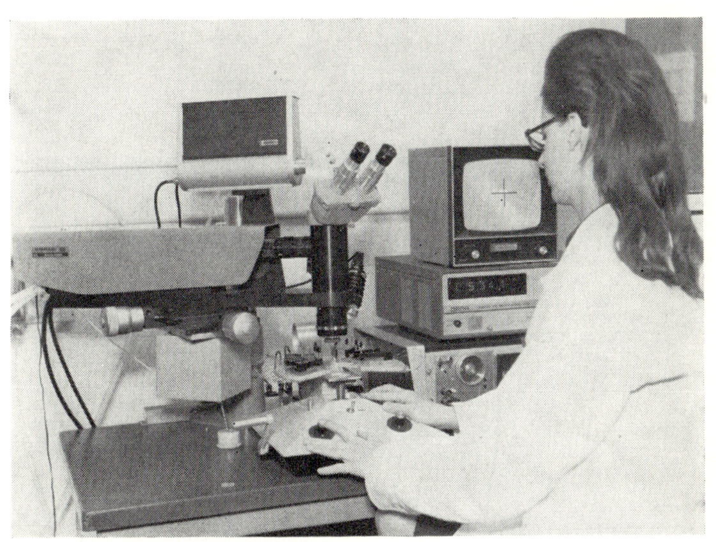

Plate 10.3. Laser Trimmer (Union Carbide, Korad Dept.)

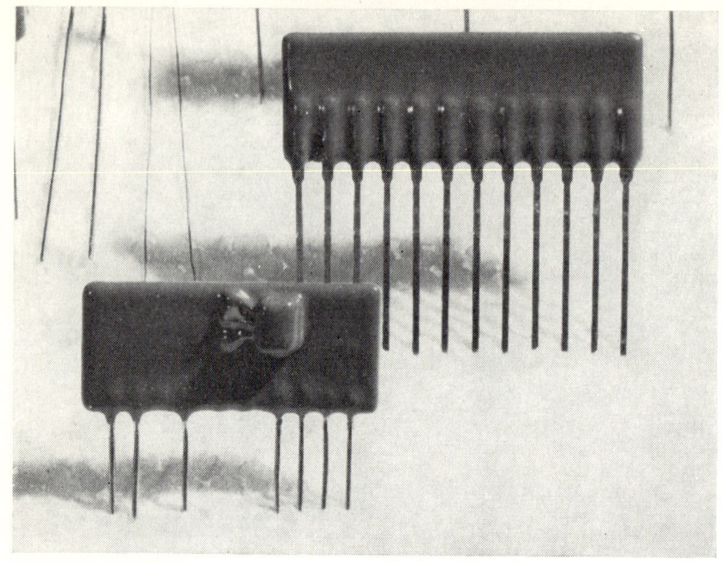

Plate 11.1. Conformal Coated Circuit (Hysol Sterling)

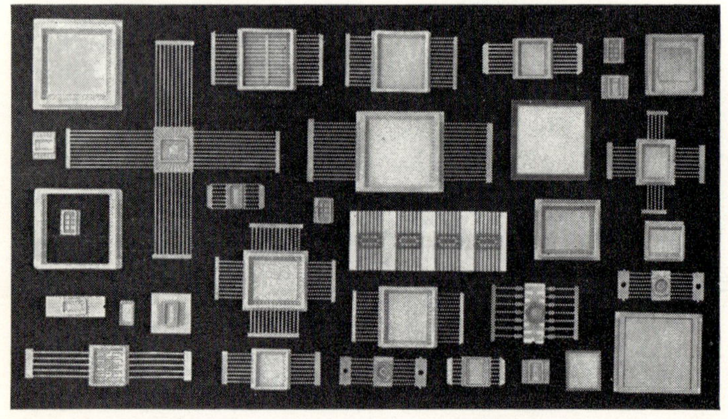

Plate 11.2. Hermetic Packages (Coors Porcelain)

formulation over previously deposited conductive lines. The pillars are then subjected to a so-called 'coining' treatment, wherein they are compressed in a steel die to give uniform height and surface area for subsequent bonding.[6]

In another approach, pillars of aluminium, copper or gold are formed on the substrate by vacuum deposition. The pillars can either be produced directly by evaporation through a mask defining the pillar areas, or they are delineated by photoetching a 'moat' near the end of each conductor pattern.[7] In yet another procedure gold coated tin-lead pillars are electrolytically deposited at the ends of conductors in the chip mounting area of the conductor pattern.[8]

In the case of the pillar-on-chip approach, a number of different metallurgical systems have been employed to produce the raised contact bump and join the chip to the conductor pattern on the substrate. Some degree of hermetic protection is often provided by passivation of the active areas of the chip with a layer of silicon monoxide or glass.

In the so-called SLT process, nickel-plated copper balls are produced on chromium–copper–gold contact pads on the chip.[9] The latter is placed on a solder-coated conductor pattern having dimples formed in the appropriate positions to receive the contact bumps of the device. The solder is then reflowed in an oven to bond the assembly.

In a further development of this method, the copper ball is replaced with a malleable pad of vacuum deposited tin–lead solder. In addition, tinnable contact areas on the substrate are surrounded by non-tinnable screened glass barriers which, during reflow soldering, restrict the flow of molten solder on the interconnection pattern so that surface tension retains the chip suspended above the substrate. This so-called 'controlled collapse' technique is simpler metallurgically, and is claimed to produce more reliable ductile joints and to be suitable for use with relatively complex chips.[10]

An alternative procedure to the use of screened glass barriers is that of depositing solderable contact areas over a non-solderable interconnection pattern, using for example palladium–silver and gold compositions, respectively (see Chapter 4). The molten solder is in this way restricted to the tinnable areas. The problem of accurate registration of the two materials is, however, more acute in this case. A further method of controlling the collapse of solder pad devices on reflowing utilises a curtain of cold, inert gas such as nitrogen around the periphery of the device to limit the flow of the molten solder.[11]

The solder reflow of, for instance, copper-ball devices has been shown to provide a reliable and effective method for automation in production.[12] Since the ductile soft-solder joints can take up considerable stresses during thermal cycling, strains are thereby reduced with resultant enhanced reliability. Moreover, some degree of non-linearity of the pads or lands to which they are joined is permissible, since the solder is malleable and capable of some distortion during the reflow soldering process.

Solder reflow bonding of chips having solder ball bumps can be carried out using a jet of heated gas, for example, of 10% hydrogen-nitrogen mixture.[13] This technique has been found to produce a consistently high bond strength and the method is capable of automation in large scale circuit production.[14]

9.1.3 BEAM LEAD DEVICES

Beam lead devices are essentially semiconductor chips having integral connecting leads of gold or aluminium which extend from the contact areas on the chip and protrude over the chip periphery in the fashion of a miniature flat pack (see Plate 9.2). In the case of the beam lead technology developed by Bell Telephone Laboratories[15,16] the chip is, in addition, passivated on the surface by a layer of silicon nitride, thus providing, at the chip level, a hermetically sealed element. Devices having from 2 to 50 or even more, beam leads have been produced.[17] A major advantage of this approach is that with suitable equipment, bonding of the device to a thick film circuit can be carried out effectively in a single operation. Improved reliability is claimed for silicon nitride passivated devices of this type and failure rates as low as 0.001% per 1 000 hours have been reported at operational temperatures of over 200 °C.[18] Lower overall manufacturing costs can be expected for assemblies incorporating beam lead devices.[19] Although the fabrication costs of a beam lead wafer are relatively high, final yields are normally better than for conventional chips, due particularly to the process of die separation which is carried out by chemical etching rather than by scribing and flexing, thus virtually eliminating the risk of broken and damaged dice. After separation, chip orientation can be maintained, so reducing the amount of handling required. Devices can then be readily tested to allow rejection of any sub-standard elements.

Devices with silicon nitride surface and junction coatings do not require costly hermetic sealing since this material acts as a barrier against contamination by ions, particularly of sodium, which will penetrate conventional oxide passivation layers.[20,21] Encapsulation can then be limited to a thin layer of a silicone resin to prevent electrolytic corrosion between the closely spaced beam leads in the presence of moisture or other contaminants, followed if required, by an additional coating of a silicone or epoxy resin for mechanical protection.

According to one system[15, 16], the beam leads consist of gold; they normally project about 0.005 in beyond the periphery of the chip and may range upwards in thickness from 0.00025 in and in width from 0.001 in. The beam lead production process is an extension of the standard planar integrated circuit fabrication technique. Following the silicon nitride passivation process, ohmic contact to the devices is made by the deposition of a platinum silicide layer. A titanium diffusion barrier is now provided, followed by electrolytic deposition of the gold beam lead material. The leads are then defined and the chips subsequently separated by chemical photoetching processes.

Alternative beam lead systems employing other metallurgical combinations are also in use. One of these utilises an all aluminium system, the beams, being produced by photoetching of a vaccum deposited aluminium layer.[22] In an extension of this system ohmic contact to the chip is again made by vaccum deposition of aluminium, followed by the formation of gold beam leads; these are laterally separated on the surface of the chip from the aluminium areas, to which they are then joined by molybdenum interconnections.[23]

To take advantage of the integrated multi-lead system offered by beam lead devices, several techniques for single operation bonding have been developed. The process is complicated by the small dimensional differences which may exist in the leads and contact pads to which bonding is carried out. Such variations may cause non-uniform deformation of the leads and hence lead to lowered reproducibility of bond quality.

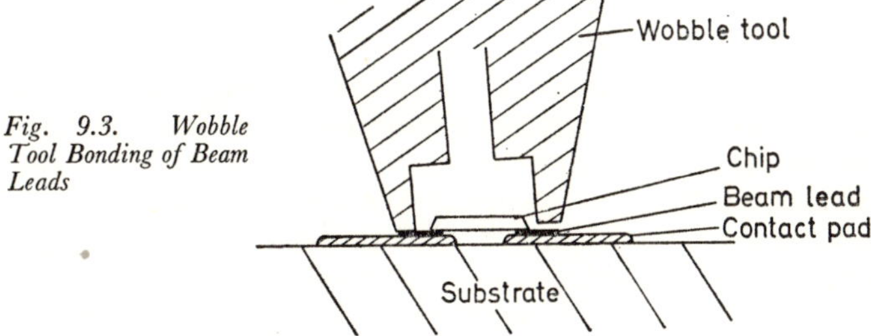

Fig. 9.3. Wobble Tool Bonding of Beam Leads

One method suitable for gold beam leads employs a thermocompression 'wobble tool,' illustrated in Figure 9.3. This comprises a rectangular section, hollow head which is rocked about its vertical axis through a small circular solid angle, so bonding each lead sequentially around the periphery of the chip. An alternative method also used to obtain the requisite 'wobble' action is to tilt the substrate table about its vertical axis so that the beam leads are compressed sequentially beneath the bonding head.

A further approach of these problems is the so called compliant bonding technique, which can be used for both gold and aluminium beam leads.[24] In this method, illustrated in Figure 9.4, a deformable or compliant medium such as an aluminium sheet of about 0.005 in thickness is interposed between the leads to be bonded and the bonding tool. When bonding, the tip is brought into contact with the compliant medium which then conforms to the topography of

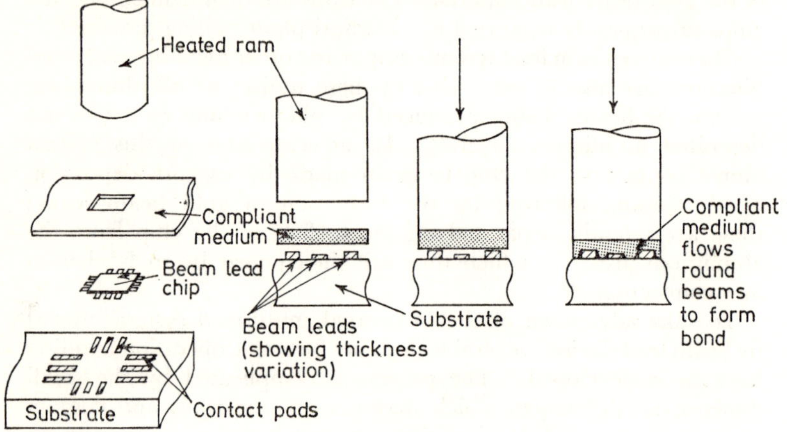

Fig. 9.4. Compliant Bonding of Beam Leads

the bond region as it transmits the bonding energy. The compliant medium also controls the deformation of the lead by transmitting a force gradient along the axis of the lead while it partially constrains the metal flow, so resulting in controlled deformation of the bond region. This is particularly important in the case of closely spaced, narrow beams (0.002 in beam width and spacing), to prevent lateral spread and the danger of electrical shorting.

As a result of the inherent flow properties of the compliant medium, the process reduces the problem of variations in tolerances significantly, at the same time controlling the extent of lead deformation and facilitating the transmission of a uniform degree of bond energy to a multiple number of leads in one process. In contrast with other solid state bonding processes which are highly dependent on the precision of the bonding tool employed, compliant bonding relies rather on the energy transmission characteristics of the compliant member for the control of the process. Using this principle, bonding may be carried out by the application of heat and pressure (that is thermocompression bonding), or of ultrasonic energy to the compliant medium, the tool in either case being flat and requiring no precision shaped tip as in other bonding techniques.

9.1.4 AUTOMATED CHIP ASSEMBLY TECHNIQUES

A number of procedures of chip bonding for use in automatic large scale assembly have been developed. In the 'chip-and-finger' technique, continuous belt lead frame configurations are employed, having 'finger' interconnections to which the chip contacts are bonded simultaneously on an automated assembly line. The lead frame, which is fabricated from unsupported etched or stamped Kovar, is then bonded to the ceramic conductor substrate in a second operation. Whilst this technique is suitable for automatic assembly and can make use of a wide range of standard semiconductor chips, it requires a relatively large substrate interconnection area and may lead to intermetallic alloy formation problems, unless aluminium interconnections are used.

In the place of unsupported metal lead frames, an array of etched metal conductors on a backing of polyester or polyimide film may be employed. In one approach, some 72 tinned copper fingers extending beyond a square hole cut in the plastic are soldered simultaneously to gold plated bump contacts on an IC chip positioned in the hole, the chip array being then soldered to the substrate interconnection pattern.[25] In another technique, the chip is first back-bonded into a recess in the substrate and interconnected by means of a similar film based, etched configuration. In some cases the film may subsequently be removed. This technique, which utilises standard chips, is again suited to automatic production procedures and is capable of yielding a higher density of active devices than does the unsupported lead frame, 'chip-and-finger' method.

In another approach, beam leads are formed on an alumina or polyimide substrate by depositing the interconnection pattern on one face of the plate and etching a window from the rear so that overhanging beam leads are produced. The chip is then inserted in the window from the rear, active face uppermost, and is connected to the beam leads in one thermocompression or ultrasonic bonding operation. This technique again makes use of standard chip devices. An additional heat sink may be bonded to the rear of the chip to improve thermal dissipation.[26]

According to yet a further approach suitable for large scale production, the chips are back-bonded in recesses in the ceramic

Table 9.2

COMPARISON OF DEVICE CHARACTERISTICS

Characteristics	*Chip and Wire*	*Flip Chip*		*Beam Lead*	*Ceramic mounted (LID)*	*Chip and Finger*	*Beam Lead Substrate*	*Recessed chip, deposited connectors*
		bumps on substrate	*bumps on chip*					
Cost Chip	low	low	medium	high	high	low	low	low
Cost Attachment	high	low	low	low	medium	low	low	low
Reliability	low	high	high	high	medium	high	high	high
Circuit Density	medium	high	high	high	low	medium	high	high
Thermal dissipation	high	medium	medium	medium	low	high	high	high
Limiting operating frequency GHz	0.5	1	1	> 10	0.5	10	> 10	> 10
Ease of Inspection	good	poor	poor	good	poor	good	good	good
Yield	low	medium	medium	high	low	high	high	high
Availability of chips	good	good	poor	poor at present	poor	good	good	good

substrate. The spaces between the edges of the chip and the substrate are then filled with a dielectric plastic or inorganic material, selected to give matching thermal expansion, thus forming a flush surface with the upper faces of the chips. Interconnections are then deposited simultaneously by vacuum evaporation or by electroplating.[27]

The so called STD process is a multilayer technique for the large scale batch production of hybrid circuits utilising standard IC chip devices.[28] A first level of a thick film conductor pattern is formed on the substrate, which may be alumina, beryllia or a ferrite ceramic. 'Mesas' or 'vias' are then deposited to carry the interconnection pattern from the lower level to an upper chip pad level, and IC chips are eutectically back-bonded to the substrate. A thin layer of thermoplastic fluorinated ethylene propylene co-polymer (FEP) is now bonded over the entire assembly by application of heat and pressure, and holes are etched through the plastic to the 'mesas' and chip pads. These are interconnected simultaneously by application of a second, or upper, level of metallisation on the upper surface of the FEP layer. The bonds formed with the IC contacts are of the diffusion type.

9.1.5 COMPARISON OF CHIP JOINING TECHNIQUES

A comparative summary of the salient characteristics of the major chip joining techniques discussed is given in Table 9.2. Whilst the chip-and-wire approach is attractive for the small scale production of hybrid assemblies for low frequency operation, the low chip cost and flexibility of layout are counteracted by high assembly costs and relatively poor reliability. The latter may be improved by the use of flip chip and beam lead devices. For applications requiring large scale assembly with high reliability and high speed operation, one of the newer batch processes may ultimately prove to be most suitable.

9.2 PACKAGED SEMICONDUCTOR DEVICES

In view of the difficulties of testing and assembling the bare chip devices discussed in the previous section, packaged and pre-tested semiconductor devices are often used in commercial production. To this end, a growing number of devices have become available specifically for hybrid circuit applications. Amongst these are miniature plastic encapsulated devices, dual in-line packages (DIPs), flat packs and leadless inverted devices (LIDs). They are, in general, rather more expensive than the simpler types of non-encapsulated standard chip devices, but they offer the circuit manufacturer the advantages of pretested components, together with relative ease of handling and assembly. The initial high cost of the devices is thereby offset by the resulting higher overall yield rates attainable.

In most cases, bonding of the devices to the thick film circuit is accomplished by relatively simple reflow soldering techniques employing hot gas, oven, soldering iron or focussed infra-red heating methods. In reflow soldering, one or both of the component contacts to be joined are pretinned; they are then brought together and heated to melt the solder and so form the bond on subsequent cooling. The solder used must be compatible with both the device contact and thick film conductor materials. For normal application the solder employed is a tin-lead alloy containing a small proportion of silver (see Chapter 4); however, some special requirements may dictate the use of relatively low or high melting point alloys.

One conventional method of reflow soldering is the hot gas technique, in which a jet of heated gas, typically nitrogen, is directed onto the areas to be joined.[29] Plate 9.3 shows a semi-automatic equipment for the location and attachment of plastic encapsulated devices, flat packs, dual in-line packages etc. The contact areas are automatically scanned beneath two fine jets of heated nitrogen to solder the connecting leads on either side of the device; alternatively, fan shaped jets are used to solder the leads. The flow time and flow rate of the gas are controlled at preset values. Location of the device in the circuit pattern is effected by means of an automatically operated vacuum 'finger'. Close control over soldering conditions with resultant high bond reliability is achieved in this way. The protection conferred by the inert gas and absence of contact with soldering tools ensures clean joints, and the accurate localisation of heat afforded in this way safeguards circuit components adjacent to the bond area. If oxygen-free nitrogen is used as the heating gas, it is possible under some conditions to dispense with the use of soldering fluxes; alternatively, small quantities of a non corrosive flux are often used to obtain more rapid flow of the solder.

A manually operated hot gas soldering machine is shown in Plate 9.4; the substrate, carried on a heated substage, is here positioned with the use of an X-Y micromanipulator. Multi-headed hot gas soldering equipments are sometimes used for the reflow soldering of pretinned components on a continuous conveyor belt system.

Direct contact, heated tool soldering or oven soldering are also used in production. In the former case, for small scale production the leads are individually soldered by an operator using a temperature controlled miniature soldering iron, normally with some form of magnifying viewing aid. Alternatively, automatically operated soldering tools are employed, positioning of the component on the substrate being effected in a similar manner to that employed in the automatic hot gas soldering machines described. Hand soldering naturally tends to result in poorer reproducibility and therefore lower reliability of the joint than is attainable by, for example, the non contact hot gas methods with their better control of soldering conditions.

In the case of oven soldering, the pre-tinned components are manually located on the substrate using a flux which temporarily retains the component in position. The solder is then reflowed by heating the substrates in a batch oven, or by passing them through a continuous oven on a conveyor belt.

Yet a further method of reflow soldering utilises infra-red radiation from a tungsten filament or halogen-quartz lamp, focussed onto the work by a parabolic reflector. Although linear focal areas can be obtained with appropriate reflectors, a circular focal spot of less than 0.1 in diameter cannot easily be obtained, and these methods are therefore applicable mainly to the larger types of component such as dual in-line packages and flat packs. One of the difficulties encountered in focussed infra-red systems is the necessarily small distance between the work and the base of the reflector. This imposes restrictions on the use of device manipulators or other automatic handling mechanisms.

9.2.1 PLASTIC ENCAPSULATED DEVICES

This type comprises miniature encapsulated diodes and transistors (see Plate 9.5), and dual in-line packages. Both employ similar production processes utilising a lead frame technique. The semiconductor chips are back-bonded to a lead frame and connection between the chip contact areas and the frame are made by one of the standard wire bonding techniques, for example, by thermocompression bonding of gold wires. The lead frame typically comprises a metal alloy such as Kovar which is gold plated to permit bonding of the connecting wires from the semiconductor chip.

The bonded chip/lead frame assembly is then encapsulated in an epoxy resin by a transfer moulding process. In this, a multiple lead frame assembly is positioned in a multi-cavity mould which is then filled with the liquid encapsulant material and cured at elevated temperature.

Prior to transferring the resin into the cavities in the mould, it is heated to liquefy it and is transferred at low pressure to obviate the danger of damage to the fine wire interconnections between the chips and lead frames. After curing, the encapsulated assembly is removed from the mould and separated into individual devices by cropping the framework of the lead assembly. At the same time the leads are bent, if required, so that the areas which are to contact the substrate extend below the package, so permitting subsequent intimate contact with the bonding pads without interference by the lower surface of the package. If the device is intended for attachment by reflow soldering, the leads are coated with a layer of solder.

In the fabrication of plastic encapsulated devices, considerable care is exercised to prevent interaction between contaminant materials in the resin and the active chip surface. The latter is passivated with a vacuum deposited glassy material during fabrication and may additionally be protected by a thin film of an ultra pure silicone varnish. The resin encapsulant is, moreover selected for freedom from inorganic salts, ions or gases, care being taken when making up the resin to ensure precise stoichiometry of the resin and hardener constituents to meet these requirements.

9.2.2 LEADLESS INVERTED DEVICE PACKAGES (LID)

The LID device represents a compromise between the non-encapsulated and fully encapsulated forms. The hybrid circuit manufacturer utilises a device which is not fully encapsulated but which has been tested by the semiconductor manufacturer and is relatively easy to handle and to bond to the thick film substrate.

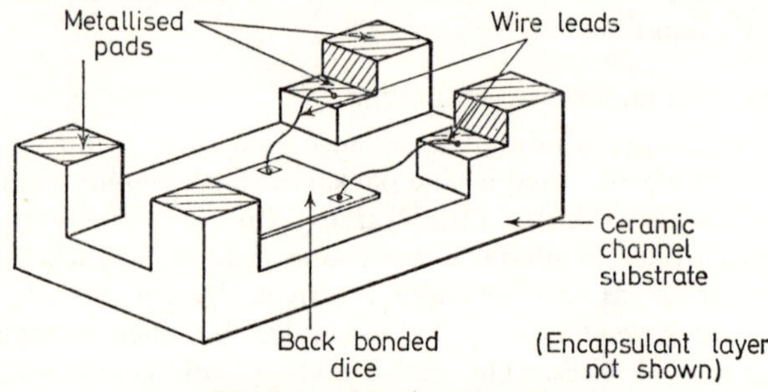

Fig. 9.5. LID Type Transistor

In this configuration, the semiconductor dice is back-bonded to a channelled ceramic chip provided with stepped metallised contact pads, as illustrated in Figure 9.5. Wire leads are connected to the contact pads on the device and the lower metallised steps on the ceramic holder. The chip and interconnections are then encapsulated with a protective resin covering.

The device is connected to the thick film conductor pattern by a solder bonding technique in a similar manner to that used for the fully plastic encapsulated devices.

An alternative method of attachment of ceramic mounted chips is to cement the rear face of the carrier to the substrate, and to connect wire leads between the carrier pads and the substrate contact areas by thermocompression or ultrasonic bonding.

LID devices are somewhat more expensive than the conventional plastic encapsulated devices; moreover, the encapsulant provides poorer environmental protection than in the latter case.

9.2.3 HERMETICALLY SEALED TYPES

The hermetically sealed class of devices includes flat packs, dual in-line and T05-type configurations. In the first two types, the chip is back-bonded to a lead frame and in the last case, to a circular header. Fine wire interconnections are provided between the chip contact areas and the lead frame or the header. The packages are then hermetically sealed employing glass to metal or ceramic to metal seals by soldering, brazing or by welding (see Chapter 11).

The flat pack and the DIP configurations employ rectangular section leads, which after suitable forming permit attachment to thick film substrates by any of the solder reflow methods already described. The T05-type package, however, incorporates wire connectors disposed in a circular configuration, making this package less suitable for thick film applications. Nevertheless, such types are often used in cases where a particular device is not available in any of the more suitable packaging systems, the leads normally being fed through holes in the substrate and soldered in position.

In comparison with fully hermetically sealed devices, plastic encapsulated devices have relatively poor resistance to moisture and tests have shown that hermetically sealed devices are superior to plastic encapsulated devices under humidity test conditions by a factor of 2 to 1.[30] However, development of improved methods of surface passivation of the active device, in conjunction with more efficient plastic sealing techniques, are likely to increase the field of application of plastic encapsulated devices.

9.3 CAPACITORS

Capacitor chips of the monolithic ceramic or sintered tantalum types can be employed to overcome the capacitance/area limitations inherent in screened and fired capacitors. They also obviate the need to employ special high K substrates to attain higher capacitance levels (see Chapter 6). For example, capacitance values up to the order of 3 μF can be obtained in a chip area of about 0.6 × 0.8 in in the case of a ceramic chip capacitor, or values as high as 68 μF are attainable in tantalum chip capacitors having dimensions of the order of 0.2 × 0.3 in.

9.3.1 MONOLITHIC CERAMIC CHIPS

A monolithic ceramic capacitor comprises a compressed stack of thin dielectric layers interleaved with electrode layers. This configuration yields a maximum capacitance for a given sized package by giving a greater electrode area at the expense of increased height (see Figure 9.6).

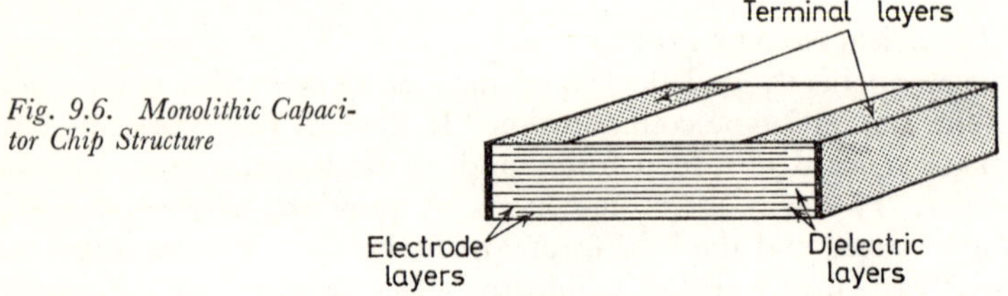

Fig. 9.6. Monolithic Capacitor Chip Structure

The fabrication of these devices involves the casting of a ceramic dielectric slip into thin flexible strips and screen printing these with a noble metal paste, usually containing platinum. The strips are stacked to give the required number of layers and are fed into a tool which punches and consolidates the chips under pressure. The compressed chips are ejected from the tool and parallel connections with the edges of the internal electrode layers are made by applying a conductor composition. The chips are then fired to give a monolithic block having the appearance and physical properties of a single piece of ceramic material. The sintering temperature is generally in the range of 1 300–1 450 °C depending on the type of dielectric material in use. By this technique some 25 or more dielectric layers, each having a thickness of about 0.001 in, can be formed, the thickness of each electrode layer being generally of the order of 0.0001 in.

The dielectric materials employed are similar to those used in conventional capacitors and in screenable dielectric glazes (see Chapter 6). In general, two types are obtainable, viz, the high stability NPO (negative positive zero) and the high K types.

The NPO type is based on materials having a linear temperature-capacitance curve of nominal zero slope, and a permittivity of about 30. The temperature coefficient of capacitance of this type is generally in the region of $\pm$ 30 ppm/°C over the temperature range of –55–125 °C. Capacitance values range from about 1.0 pF to 0.10 μF at a voltage rating of 50 volts dc; higher voltage ratings can be obtained, but at the expense of a reduction in the maximum operating temperature of about 85 °C.

For less stringent stability and temperature requirements high K capacitors based on barium titanate materials are available. These have a higher capacitance/volume characteristic than the NPO type and can be obtained in values up to about 3 μF in a size of about 0.6×0.8 in. In this case, capacitance density approaches 100 μF/in^3.

Temperature coefficients of capacitance vary with the permittivity of the dielectric material; typical values are ±15% at a K of 1 700, and +20 to −85% at a K of 7 000 over the temperature range from −55 –125 °C.

Attachment of ceramic capacitor chips is normally by reflow soldering, and for this purpose the metallised edges are supplied in pretinned form. They are, however, also available with wire or ribbon leads, or in some cases with metal bump contacts for application by thermocompression or ultrasonic bonding techniques. Other types are available in which the electrodes are on opposite faces of the chip. The lower pad is here solder bonded and the top is connected by means of a separate wire to an adjacent conductor pad.

9.3.2 SINTERED TANTALUM CHIPS

The tantalum chip capacitor comprises a sintered porous tantalum slug which is internally anodised by an electrolytic process. A cathodic layer of manganese dioxide is deposited on the internal surfaces of the porous slug by impregnating it with a solution of manganese nitrate, which is subsequently pyrolised to convert it to the dioxide. Connection to the cathode is made by depositing a graphite layer, followed by a silver layer which is subsequently tinned with solder. The anode connection consists of a nickel tab, part of which is incorporated within the body of the anode slug during its formation at high pressure, whilst the remaining section is subsequently coated with a solder layer.

Capacitance values are obtainable within the range of 0.1–68 μF over a working voltage of 4–50 V dc. Temperature coefficients of capacitance are within the range of +15% at +125 °C to −12% at −55 °C.[31]

9.4 MISCELLANEOUS COMPONENTS

9.4.1 RESISTORS

While thick film circuits normally comprise printed and fired resistors, incorporation of miniature discrete resistors may on occasion be warranted in order to obtain the most efficient circuit layout, or a reduction in production costs. Such instances may arise, for example, when only one or two resistors, having values which would necessitate the use of different pastes and hence, additional screening operations for each, would be required.

Chip resistors are available in thick film, thin film and cermet form. The first two are produced respectively by conventional screening and firing, and by vacuum evaporation techniques, usually on ceramic substrates. Cermet resistors are formed by pressing and sintering mixtures of powdered resistive and insulant glass materials to form a monolithic body. As with chip capacitors, chip resistors are normally provided with metallised contact areas suitable for reflow solder attachment to the hybrid circuit, although some types are available having gold beam lead terminations for attachment by thermocompression bonding.

Thick film and cermet chip resistors are normally available in the range of from 10 ohms to 15 MΩ in 5% and 10% standard tolerances. TCRs are in the range of 50 to 300 ppm/°C with tracking from 25 to 100 ppm/°C; resistance stability is of the order of 0.5% over 1 000 hrs at 150 °C. Thin film resistors range in value to 1 MΩ, with TCRs of better than 75 ppm/°C and tracking of 2 to 5 ppm/°C. Stability is normally better than 0.05% over 1 000 hrs at 125 °C.

Discrete chip, short-circuit bridges for use as interconnecting crossovers are also sometimes used. These are similar in appearance and method of attachment to resistor chips.

9.4.2 INDUCTORS

Three types of discrete inductor are employed in thick film hybrid circuits, namely thin film spiral, wound toroid, and wound rod core inductors.

Alumina based, thin film spiral types yield values of from about 30 to 230 nH with Q values of 20–30, being used typically in UHF and microwave circuitry. Wound toroids having a diameter of about 0.08 in yield inductances of up to 100 μH. This type, possessing high permeability particularly at low frequencies, is used mainly for pulse and broadband transformers. Rod cores are wound on a cylindrical bobbin which may contain an adjustable iron core to permit tuning. Diameters range up to about 0.08 in, inductors of this size and 0.19 in long being available in values from 0.06–2 000 μH. Fixed rod cores of similar size may have inductances of up to about 3 300 μH.

Rod core inductors have restricted magnetic fields and good temperature stability, being used in RF and IF amplifiers, discriminators, tuners and modulators. Microinductors of the types described are generally connected to the hybrid circuit by wire soldering methods.

REFERENCES

[1] Hansen, M., *Constitution of Binary Alloys*, McGraw Hill, (1958).

[2] Phillips, L. S., 'Thermocompression Bonding to Thin Film Microcircuits', *British Communications and Electronics*, (June, 1963).

[3] Phillips, L. S., 'Microbonding Techniques', *Microelectronics and Reliability*, Vol. 5, 197–201, (1966).

[4] Baker, D., and Bryan, I. E., 'An Improved Form of Thermo-Compression Bond', *British Journal of Applied Physics*, **16,** 865–872, (1965).

[5] Philofsky, E., 'Intermetallic Formation in Gold–Aluminium Systems', *Solid State Electronics*, **13,** 1391–1399, (1970).

[6] Matcovitch, T. J., Coren, R. L., Keleman, D. G., and Galli, R. J., 'Ultrasonic Bonding of ICs to Thick Film Pedestals', *Microelectronics*, 14–20. (April, 1970).
[7] Curran, J. E., 'A Flip Chip Interconnection and Packaging System', *Proceedings, Inter NEPCON*, Brighton, (1969).
[8] Bingham, K. C., and Gurler, Y., 'Ultrasonic Chip Bonding Techniques', *The Radio and Electronic Engineer*, **37,** (1969).
[9] Davis, E. M., et al., 'Solid Logic Technology—Versatile High Performance Microelectronics', *IBM Journal of Research and Development*, 102–114, (May 1969).
[10] Norris, K. C., and Landzberg, A. H., 'Reliability of Controlled Chip Collapse Interconnections', *IBM Journal of Research and Development*, **13,** (1969).
[11] Wagner, S., and Walker, M., 'Low Cost Integrated Circuit Techniques', *US Research Report AD 635183*, (July, 1966).
[12] Miller, L. F., 'Joining Semiconductor Devices with Ductile Pads', *Microelectronics*, 24–29, (April, 1969).
[13] Evison, P. J., 'The Face Bonding of Semiconductor Chips', *Microelectronics and Reliability*, **5,** 209–212 (1966).
[14] Khambatta, A. F., and Castle, P. F., 'Face Bonded Chips and Thick Film Compatibility', *Electronic Packaging and Production*, 33–38, (September–October, 1970).
[15] Lepselter, M. P., 'Beam Lead Technology', *Bell System Technical Journal*, 233–253, (February, 1966).
[16] Lepselter, M. P., 'Beam Lead Sealed Junction Technology', *Bell Laboratories Record*, (October/November, 1966).
[17] Mallery, P., 'Bonding Beam Leads', *Proceedings of Conference on Microelectronic Circuits*, 1–6, (August, 1968).
[18] Peck, D. S., 'Reliability of Beam Lead Sealed Junction Devices', *Proceedings of Annual Symposium on Reliability*, 191–201, (1969).
[19] Acosta, J. R., 'Beam Lead Reliability in Systems Applications', *Proceedings of Eastern Electronics Packaging Conference*, 5.3.1.–5.3.9., (June, 1970).
[20] Eleftherion, M. P., 'Assembling Beam Lead Sealed Junction Integrated Circuit Packages', *Solid State Technology*, 27–36, (August, 1968).
[21] Hutton, M. E., 'Beam Lead Technology', *Microelectronics*, 32–35, (April, 1970).
[22] Neuhuys, T., 'Integrated Aluminium Connections', *Proceedings IEEE Electronic Components Conference*, 46–51, (1969).
[23] Gelshing, R. J., and Van Steensel, K., 'A Multilayer Interconnection System with Gold Beam Leads', *Microelectronics and Reliability*, **8,** 325–329, (1969).
[24] Coucoulas, A., 'Compliant Bonding', *Proceedings, IEEE 20th Electronic Components Conference*, 380–389, (May, 1970).
[25] 'Integrated Circuits Will be Coming on Reels of Film', *Electronics Weekly*, (January 27th, 1971).
[26] Bachner, F. J., Cohen, R. A., and McMahon, R. E., 'Rigid and Non-Rigid Beam Lead Substrates', *Solid State Technology*, 62–66, (August, 1970).
[27] 'Facing up to the Chip', *Electronics Review*, (July, 1967).
[28] Clark, R. J., and Lunden, J. W., 'The Application of the STD Process to Hybrid Microelectronics', *Proceedings, IEEE 20th Electronic Components Conference*, 216–235, (May, 1970).
[29] Phillips, L. S., 'Hot Gas Soldering Applied to Electronic Assembly and Component Production', *Proceedings, Inter NEPCON*, Brighton, (October, 1971).
[30] Reich, B., 'Plastic Semiconductor Devices and Integrated Circuits for Military Applications', *Solid State Technology*, 53–56, (January, 1970).
[31] Maguire, D., 'Improved Tantalum Chip Capacators for Hybrid Circuits', *Proceedings, IEEE, Electronics Component Conference*, Washington, 211–217, (1969).

10

Trimming and Test Procedures

Test and trimming procedures are applied to thick film circuits to ensure that their ultimate electrical characteristics conform to specification. Trimming of screened resistors and capacitors to specified values is normally necessary, since, in the deposition process, these cannot be produced consistently to tolerances of better than 10 to 20%. The tolerance attained initially depends on the type of composition utilized and the degree of control exercised over the deposition and firing conditions, as already discussed in Chapters 5 and 6.

Trimming may be carried out on a partly completed circuit having printed passive components only, and in this case individual resistors and capacitors will be separately adjusted to the specified values. Alternatively, a virtually complete hybrid assembly incorporating both printed passive components and subsequently added active devices, may be tested and adjusted to ensure that it functions according to a set electrical specification. In the latter case, functional, or 'dynamic trimming', of one or two critical resistors only is normally required, these being adjusted to bring the circuit parameters to within the specified limits. Trimming is a time consuming, and therefore relatively costly, process and accordingly significant cost savings can be made if the circuit design is such that the functional trimming technique can be utilized.

Testing, whether of individual resistors or capacitors, of active semiconductor devices subsequent to bonding to the circuit, or of the completed circuit system, necessitates the positioning of fine probes, connected to an external test facility, upon the appropriate contact points in the circuit.

Trimming is effected by removing small areas of the printed component, usually by air abrasive or laser techniques. In the former process, which is well established in commercial practice, a small air jet carrying fine particles of an abrasive powder is directed at the component, part of the surface being thereby removed by what is virtually a miniature form of sand blasting. In the latter, relatively new process, a laser beam is directed onto the component, a portion of which is removed by vaporisation. Equipments are available in which component values can be adjusted automatically to preset limits using either technique.

10.1 TEST METHODS

The testing of circuit component characteristics necessitates that electrical contact is made with appropriate conductor areas on the substrate. For this purpose, fine probes mounted on micromanipulators to permit movement in three planes are utilised. For hybrid circuit applications, the excursion must be sufficiently wide to permit full X and Y plane coverage of the substrates, which typically may measure as much as 4 in $\times$ 4 in. In addition, means for adjusting the probe contact pressure is normally provided. A probe assembly positioned on a thick film circuit is illustrated in Plate 10.1.

The probes, which are often fabricated from tungsten carbide, may be either single-pointed or parallel, electrically insulated, twin-pointed. The latter, Kelvin type, is employed to provide separate current and voltage paths to the circuit, thus reducing errors by eliminating contact resistance effects. The parallel Kelvin probes must be closely spaced to permit positioning of probes on contact pads which may, in some applications, have widths as small as 0.01 in, with separations of similar dimensions between pads. A microscope is normally used with multiprobe systems to permit adjustment of the probes to be made relative to the contact points on the circuit when setting up for testing.

The electrical test system may range in complexity from a relatively simple resistance monitor to a sophisticated computer system. The latter automatically controls the movement and functioning of a complex multihead trimming assembly in which a number of resistors are simultaneously trimmed to individually specified values.

10.2 TRIMMING METHODS

10.2.1 AIR ABRASIVE TRIMMING

In the air abrasive process a miniature jet of air carrying a fine abrasive powder, normally alumina of 10 to 50 microns particle size, is directed onto the resistor through a fine nozzle. The jet is traversed across the surface of the film which is thereby selectively removed, so causing an increase in the resistance value of the film. The process does not generate a significant temperature rise, is shock free, and providing the jet pressure is suitably controlled, does not damage the underlying substrate. The minimum width of cut is approximately 0.005 in, while wider cuts can be produced by utilising broader nozzles.

The major drawback of air abrasive systems is undoubtedly the risk of 'overspray', that is, coating and abrasion of areas adjacent to those subjected to trimming, due to the lateral drift of abrasive powder which tends to occur in spite of the use of efficient dust extraction systems. Another difficulty is that of wear of jet nozzles and mechanical working parts of the trimming appliance by the abrasive powder.

The effects of overspray on neighbouring components can to some extent be reduced by appropriate orientation of the substrate, so that the body of the component being trimmed acts as a protective shield for the surrounding areas. Surface abrasion of resistors by overspray can be eliminated if they are provided with a printed and fired protective glaze. Such a glaze is also desirable for the stabilization of palladium–silver resistors and thus serves a dual purpose in this case. However, such stabilisation is not required for the non reactive ruthenium oxide and precious metal type resistors, and application of glazes for overspray protection will here result in additional production costs. The effects of overspray may be more serious in the case of functional trimming of circuits incorporating unprotected semiconductor chips, and care must therefore be taken to avoid damage due to the drift of abrasive powder.

During air abrasive trimming, the intrinsically protective vitreous layer formed when thick film resistors are fired is abraded along the edges of the cut, thus exposing the resistive constituent of the resistor and destroying the hermetic sealing effect. In the case of the reactive palladium–silver resistors, this will tend to result in a somewhat increased ultimate drift rate in the trimmed resistors. On the other hand, such exposure has an insignificant effect on drift rates in the case of films having nonreactive ruthenium oxide or precious metal resistive constituents.

A wide range of air abrasive systems, from manually operated to fully automatic types, is available. In production facilities, resistor trimming is generally accomplished automatically, the air abrasive nozzle being moved across the resistor under the control of an automatic system incorporating a precision bridge.[1] The changing value of the resistor under trim is continuously monitored by the bridge which checks that the initial resistance value is within acceptable limits, signals the trimming equipment to commence operation and removes the trim signal when a preset value is attained. After a time delay, the resistor is generally rechecked by the bridge for final value and, if within the programmed tolerance limits, is accepted.

The interval between the stop command and cessation of trimming is of the order of several milliseconds, since abrasive particles continue to impinge upon the resistor after shut-down. This results in an overshoot in the value change, which however is normally anticipated and allowed for by the monitoring/control system[2]. Air abrasive systems are capable of trimming to within ±0.1% of specified resistance value at trim speeds of 0.15 to 1.5 in per minute. The accuracy decreases to ±0.3% at trim speeds of about 6 in per minute, whilst for accuracies of 0.5% or less, considerable increases in trim speed are possible.[3] Since the ultimate stability of a printed resistor depends on the type of composition utilised, ranging from 0.1% to 1% or above, it is often unnecessary to trim to the highest accuracies possible, and hence the higher trimming speeds may be employed. It has been reported that accuracy of trimming may be affected by the build up of static charges on the substrate. This effect can, however, be overcome by raising the moisture content of the carrier gas from the normal order of about 25 ppm to 600–700 ppm[4].

In the case of semi-automatic trimmers, in which loading and unloading of the circuit plates is carried out manually at one station while trimming takes place automatically at a second station, throughputs of 1 000 resistors per hour can be obtained for trim times of about one second. Considerable increases in production rates can be attained by automatic trimmers incorporating rotary substrate-feed tables, and multiple abrasive jets each associated with separate resistance monitoring circuits and trim control systems. This enables a number of resistors on one substrate to be trimmed simultaneously. For example, one commercial trimmer system incorporating seven heads and rotary feed tables can, with two seconds trim times, produce 12 600 trims/hr. Plate 10.2 illustrates an abrasive trimmer incorporating an 8 station rotary table with 4 preset trimming nozzles and associated monitor/control systems.

10.2.2 LASER TRIMMING

In this technique, a thin path is cut through the resistor by vaporising the material with a focussed laser beam. The glass constituent of the remaining resistive film melts along the edges of the cut, thus resealing the resistor and hence, in contrast to the air abrasive technique, reprotects the resistive constituent of the film from environmental effects. This feature is of particular significance in the case of some of the palladium–silver resistor types[3].

Laser trimming of thick film capacitors, however, may provide some difficulties due to the edge-fusion effect. There is here a tendency to cause short circuiting between the conductor layers, or reaction between the dielectric and conductor layers may result, with a consequent decrease in permittivity of the former. For this reason abrasive trimming is normally used for the adjustment of capacitors.

Overspray effects are, of course, eliminated in the case of laser trimming; any contamination being limited to redeposition of small quantities of vaporised material. Some laser equipments utilise a fine jet of inert gas to blow the vaporised material away, although this is mainly intended to prevent deposition on the focussing lens rather than on the circuit. The absence of overspray effects makes the laser technique attractive for functional trimming since there is virtually no danger of damage to unprotected semiconductor devices on the substrate.

The high power, high repetition rate lasers now available, coupled with a capability for arresting the trimming operation within pulse times of the order of microseconds, makes laser trimming an inherently rapid system in production. Electronic, rather than mechanical means of shut-down, together with electronic beam scanning, allow the systems designer to take full advantage of the high rate of material removal by the laser, whilst maintaining the necessary accuracy in the attainment of final trim values.

Minimum spot diameters as small as 0.001 in are attainable in thick film laser trimmers, with the possibility of increasing this to about 0.01 in where required, defocussing the spot.

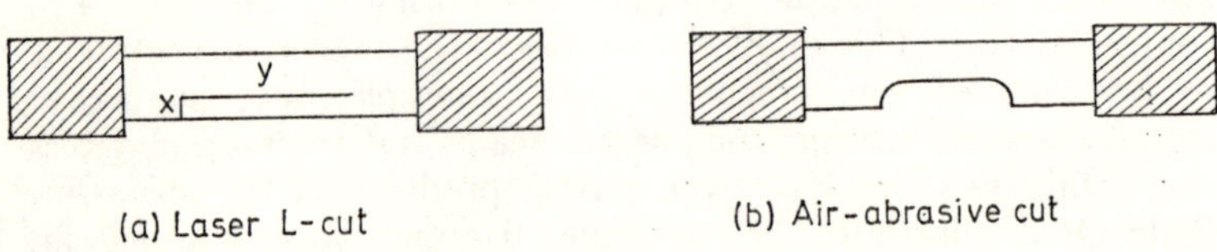

(a) Laser L-cut

(b) Air-abrasive cut

Fig. 10.1. Trim Configuration

By producing L shaped cuts, as shown in Figure 10.1 (a), it is possible to minimise the reduction in power dissipation of the resistor due to trimming. The first cut, x, is made perpendicular to the length of the resistor, this being followed by a second cut, y, parallel to the length. By making x as short, and y as long as possible, the maximum cross sectional area, and hence, ultimate power dissipation will be obtained. A computer is used to determine the optimum ratio between the two cuts for each resistor. The L type cut can be compared with that produced by abrasive trimming, illustrated in Figure 10.1 (b) which, while producing the same percentage resistance change, would result in a lower power dissipation capability of the final resistor[5].

A wide range of commercial trimmers is available, from simple manual controlled types to complex computer controlled systems, capable of trimming 7 200 resistors per hour with a single laser head (Plate 10.3).

Various techniques have been utilised for providing relative motion between the substrate and the laser beam during trimming. Many commercial systems employ motorised X–Y motion of the substrate table, together with the monitoring probes. Alternatively in some systems the laser head is moved relative to the stationary substrate, whilst in others both the head and the circuit are fixed and the laser beam is deflected by means of two mirrors attached to an X-Y table. Precautions are, of course, taken to prevent laser radiation damage to operators' eyes, filters or closed circuit television cameras being employed.

Two types of laser are in common use for thick film trimming, namely continuously pumped YAG (Neodymium doped Yttrium Aluminium Garnet), or CO_2 repetitively Q-switched, to provide 5–10 kHz pulse repetition frequencies required for high speed trimming. The CO_2 lasers are normally less expensive than YAG lasers and do not require external gas supplies for cooling. However, YAG lasers pulse more rapidly and yield smaller spot sizes, so making them inherently more accurate for trimming purposes. The YAG type tends to be more sensitive to vertical variations in the substrate, thus posing a potential problem in the case of relatively large substrates[6].

10.2.3 SCRIBE AND PULSE TRIMMING METHODS

In addition to the two main methods of trimming already discussed, scribe trimming and pulse trimming have also been employed. Scribe trimming is similar to abrasive trimming in that it is accomplished mechanically by removal of a portion of the resistive film to give an increase in resistance. The tool normally used is a diamond pointed probe vibrating at ultrasonic frequencies. Substrate movement, probe contact and resistance bridge measurement systems are similar to those used in air abrasive trimming. Whilst the diamond scribe wears more rapidly than an air abrasive nozzle, the technique possesses the advantage of relative cleanliness due to the absence of abrasive dust.

In the pulse trimming technique a high voltage discharge is applied to the resistor. This results in an initial decrease in resistance due to the fusing together of the conducting particles within the resistor film. If the pulse application time is prolonged, an increase in resistance will then occur due to subsequent breakdown of contact paths. It would appear, however, that there is some risk in this method of inducing increased resistance drift rates in the finished resistor on subsequent load.

REFERENCES

[1] Stone, G. B., Ironside, D. S. and Reynolds, P. H., 'Measuring Techniques for Trimming Thick Film Resistors', *Proceedings, IEEE, 20th Electronic Components Conference,* 536–546, (May, 1970).

[2] Sanders, P. J., 'Process Information for the Design of Automatic Trimming Equipment', *Proceedings of Hybrid Microelectrics Symposium, Dallas,* 197–200, (Oct., 1969).

[3] 'Hybrid Resistor Trimming; an Industry Report', *The Electronic Engineer,* 48–55, (October, 1970).

[4] Kans, J. W., 'Resistance Anomalies in Ag/PdO Thick Film Resistors', *Proceedings, IEEE, 20th Electronic Components Conference,* 109–115, (May, 1970).

[5] 'Laser Precision Adjustment of Cermet Resistors in Production', Application Note, Spacerays Inc. (August, 1968).

[6] Stone, G. B., 'Programmable Continuous Trimming: A Systems Approach to High Speed Laser Trimming of Hybrid Microcircuits', *Proceedings of Hybrid Microelectronics Symposium, Dallas,* 177–184, (October, 1969).

11

Environmental Protection

The final stage in fabrication of a thick film hybrid circuit is its encapsulation to provide environmental protection and to safeguard it from mechanical damage. For this purpose the circuit is encapsulated either in an organic plastic material or in a composite package employing such techniques as metal to glass, ceramic to glass, or metal to ceramic sealing. In general the latter type of package provides a hermetic seal at comparatively high cost, whereas plastic encapsulation is relatively inexpensive but affords somewhat less effective environmental protection.[1]

The major considerations in the selection of the packaging system are:

1. Efficiency of sealing,
2. Chemical protection,
3. Mechanical protection,
4. Applicable interconnection system,
5. Size-weight ratio,
6. Cost.

The requirement for fully hermetic protection is generally dictated by the intended application of the circuit. Thus packaged circuits for military and space applications must normally conform to a helium leak rate specification of between 2×10^{-7} and 1×10^{-8} atm cc/s, the figure depending on the package size. The packaging system selected must either hermetically seal the complete assembly, or individual hermetically sealed components can be employed and suitably interconnected. In cases in which the highest achievable reliability is required, a two level protection system is used in which hermetically sealed semiconductor devices are incorporated in hermetically sealed packages. The choice of method will depend upon considerations of economics and size-weight ratio. Whilst a fully hermetic sealed assembly will provide virtually complete environmental protection, a somewhat lower degree of protection is often adequate, and one of the various forms of plastic encapsulation may then be employed.

The type of external package connections used is largely dictated by the method of encapsulation chosen. The connections must, of course, be compatible with the overall system in respect of both the lead materials employed, and the process by which they are attached to other circuitry.

11.1 PLASTIC ENCAPSULATION

The most widely used plastics for general application in this context are epoxy and silicone resins, although other materials such as phenolic resins and even waxes can be employed for less stringent requirements. Epoxy resins are characterised by high mechanical strength, excellent electrical insulating characteristics, good adhesion and the capability of withstanding a wide range of environmental conditions.

Silicone resins on the other hand have superior heat resistance, but greater moisture transmission and lower adhesion, characteristics. Phenolic resins are relatively inexpensive but are not widely used since they release water vapour during curing and can cause corrosive effects due to the acid nature of the catalyst incorporated.

The methods of encapsulation commonly used are liquid casting, transfer moulding and fluidised bed coating.

11.1.1 LIQUID CASTING

In the liquid casting procedure, the circuits are placed in separate mould cavities and the encapsulating liquid, comprising the premixed resin and catalyst together with other ingredients such as fillers or plasticisers, is poured into the moulds to cover the circuits.

The moulds may consist of metal or plastic cases which are retained to form ultimately the outer walls of the package; the plastics may be of the phenolic, epoxy, melamine, diallyl phthalate or thermoplastic types. Alternatively, re-usable moulds may be used, the cast circuits being removed from the cavities after the resin has hardened. The moulds may here be made of metal, high density polythene, polypropylene, or from a flexible silicone rubber.

It is generally necessary that the circuits have leads issuing from one edge only so that these can emerge from the encapsulating resin through the open face of the mould. However, in the case of moulds fabricated from silicone rubber, a split mould can be employed, thus permitting leads to pass through apertures in the base of the cavities; the flexibility of the rubber results in the formation of adequate seals at each lead position.

Although the casting process is a batch method, some degree of automatic handling can be utilised. Automatic equipment may be employed to dispense the requisite quantities of resin and hardener, and the ingredients may be mechanically stirred to ensure complete and rapid mixing. Sophisticated equipments are available for continuous proportioning, mixing and dispensing of multi-component reactive resin systems. In order to obtain castings which are free from volatile materials or air voids, the mould filling process is often carried out under vacuum, preferably employing some form of automatic casting machine. After the moulds have been filled, they are transferred to an oven in which the resin is then cured.

In spite of the use of such automatic equipments, the relatively long filling and curing times involved make the liquid casting process more suitable for small batch production requirements, say of the order of some thousands of circuits per week. Alternatively, it is suitable for the production of high quality packages, where the efficient degassing of the resin resulting in this method is warranted.

11.1.2 TRANSFER MOULDING

Transfer moulding is a suitable method for the large-scale production of packaged circuits of regular shape and closely controlled dimensions. There is no restriction on the disposition of the leads as in the case of cast assemblies, and dual in-line or flat pack configurations can therefore be produced.

Circuit assemblies, usually attached to a multiple lead frame, are positioned in the lower part of a multicavity precision mould. The latter is then closed in the transfer moulding machine so that each circuit is precisely supported in its cavity. The encapsulant material, in the form of part-cured (B stage) pellets, is introduced into a container in the top of the mould. Heat and pressure are applied to the material which liquefies and flows through channels into each of the mould cavities until these are completely filled. The resin is then cured under the action of continued heat and pressure over a period of one to three minutes, the duration depending on the type of resin and size of the component. The mould is then opened automatically and the parts are removed and separated from the lead frame assembly. In many cases the curing process within the mould is considered to be sufficient, while in others curing is completed in a post-cure oven.

While the initial costs of the transfer moulding equipment and moulds are relatively high, production rates of several thousand units per hour can be achieved, so that in mass production the process is very economical.

To eliminate the necessity for coating the mould cavities with a release agent prior to each casting cycle, a release agent is sometimes incorporated in the resin itself. This, may, however, affect the adherence of the resin to the circuit itself, with consequent danger of moisture penetration along the connecting leads.

11.1.3 CONFORMAL COATING

In the conformal coating process, a relatively thin layer of protective resin covers the circuit, and as the name implies the external configuration conforms generally to that of the circuit itself. The coating is produced by dipping the heated component into a fluidised bed of powdered B stage resin which, on contact with the heated surface, melts and then adheres to the assembly. The procedure of heating the circuit and dipping in the fluidised bed is repeated a number of times, usually employing automatic equipment, to increase the coating thickness to the desired extent. Curing of the resin is subsequently completed by heating the assembly in an oven. Since the process involves dipping, it is mainly suitable for circuits having connecting leads along one edge only, as illustrated in Plate 11.1.

Conformal coatings can also be produced by dipping the circuits in a thixotropic resin. Although the technique is relatively simple and can be automated, repeated dipping is required to build up a sufficiency thick coating. Compared with fluidised bed coating, relatively long curing periods are required, and reproducibility of coating characteristics is less readily obtained.

11.1.4 INTERACTION BETWEEN RESIN AND CIRCUITRY

Consideration must be given to possible interactions between the encapsulant resin and the circuit components. Chemical contamination may occur from free ions, resulting from non stoichiometric proportioning of the resin and hardener, or from metallic salts present as impurities in fillers. Another form of interaction may arise from free gases liberated from some hardeners under the influence of heat during processing or subsequent operations. Such contaminants may be relatively harmless in the dry state but in the presence of even minute traces of moisture may result in corrosion effects, particularly in active devices. Some protection may be afforded to the latter if they are passivated with a layer of glass or silicon monoxide. Passivation with silicon nitride yields a considerably increased degree of protection against ionic contamination.[2] Some protection may also be afforded by the use of a superpure silicone resin layer deposited over the chip face.

Physical damage to the fine-wire connections of semiconductor chips may be caused during polymerisation of the resin due to stresses set up within the resin layer, particularly if adhesion of the resin to the component is good. Such stresses will be most severe in the case of liquid resin systems polymerising over a prolonged period of time. In contrast, the more rapid polymerisation of part-cured solid resins employed in transfer moulding and fluidised bed encapsulation results in a lower stress level.[3] However, a considerable reduction in the stress level of cast encapsulations can be obtained by the use of a semi-flexible resin system.[4]

11.2 HERMETIC SEALING

11.2.1 TYPES OF ASSEMBLY

The hermetic hybrid circuit package is typical of the flat-pack, dual in-line or TO5 type configuration, and comprises essentially a base, body, leads and closure. The sealing interfaces which have to be separately considered are those between the leads and body or base, and between the base, body and closure components. A number of commonly used types of package structures are illustrated in Figure 11.1, whilst Plate 11.2 illustrates the wide range of configurations and sizes available.

Lead seals are generally of the glass to metal type, in which the coefficients of expansion of the glass and lead materials are closely matched, or alternatively are selected so that the glass is in compression. The lead material is normally Kovar, a nickel–iron–cobalt alloy chosen in view of its compatibility with standard sealing glasses. The latter can be used to make up part of the body structure of the package (Plate 11.2 (a)), or as an adhesive to bond together the leads, base and body components (Figure 11.1 (b) and (c)). Alternatively they can function as an adhesive to bond together an assembly of alumina frames having preshaped channels to accommodate the leads, (Figure 11.1 (d)).

In other cases ceramic to metal seals, in which metal pins are brazed to molybdenum–manganese metallised areas on the ceramic, are utilised (Figure 11.1 (e)).

Another ceramic to metal sealing technique utilises conductors buried within the ceramic, these being brought to the surface both outside the closure area to enable external leads to be attached by brazing, and inside to enable contact with the circuit elements to be made, (Figure 11.1 (f)). The buried conductors are produced

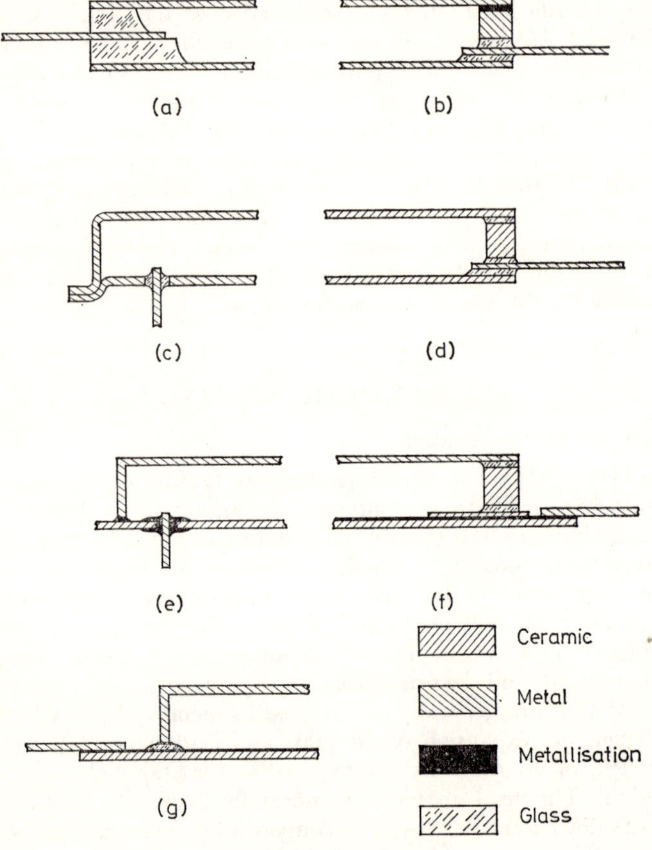

Fig. 11.1. Package Structures

by screen printing the conductor configuration on the unfired ceramic sheet, and layering the latter with a non-printed sheet; the composite is finally compressed and fired at about 1 500 °C to produce a monolithic assembly.

The base of the package may be the ceramic thick film substrate itself. This may incorporate metal contact pins brazed to metallised through-holes in the substrate, and connected to the conductor pattern by soldering (Figure 11.1 (e)). Alternatively, a layer of glass or glass/ceramic is printed and fired over the conductors in the closure area.[5] The insulating film so formed is then metallised by printing and firing a conductor composition, and a metal closure is subsequently soldered to the metallised area (Figure 11.1 (g)). A lower degree of protection is afforded by the method of attaching the closure to the ceramic base by means of an epoxy adhesive; in this case the closure may be of a plastic material.

Alternatively, the thick film substrate is mounted on the metal or ceramic base of the package utilising an adhesive, for example an epoxy resin suitably loaded to give good thermal transfer through the base. In this case the conductor lead frame is generally attached to the conductor on the substrate by soldering or brazing.

11.2.2 SEALING PROCEDURES

The requirements for satisfactory hermetic sealing may be summarised as follows:

1. The seal should afford maximum protection against ingress of vapours and moisture and be sufficiently strong to withstand the various assembly operations and handling in use,
2. The seal must be formed in an ultra dry inert atmosphere and no contaminants should be introduced into the package during sealing,
3. The temperature of the circuit elements should be kept as low as possible during sealing,
4. It should be suitable for automatic large scale assembly.

The main techniques employed for hermetic sealing are welding, brazing and glass frit sealing.[6] Welding can be carried out by projection resistance, parallel seam, and cold welding methods, illustrated in Figure 11.2 (a)–(c) respectively. Welding generally takes place in an enclosed system under a dry inert gas. After loading the circuit assemblies into the chamber, the latter is evacuated to remove air, the sealing gas being thereupon admitted.

In projection resistance welding the metallic components are brought into intimate contact in a pneumatic press, to form a uniform, narrow cross sectional contact area. A high current pulse is discharged through the copper alloy electrodes of the welder, causing simultaneous melting of the components over the whole seal area, with the formation of a hermetic seal on subsequent cooling. The components are normally of nickel–steel or a high nickel content alloy and have suitably shaped flanges on both the package and lid to allow the electrodes to make intimate contact in the weld area.

This process, which is extensively used in the production of TO5 type transistor packages is, however, mainly suitable for relatively small hybrid circuit packages on account of the high weld energy requirements in the case of larger packages.

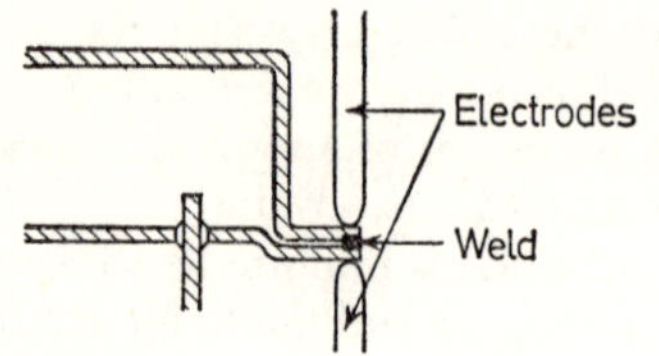

(a) Projection resistance welding

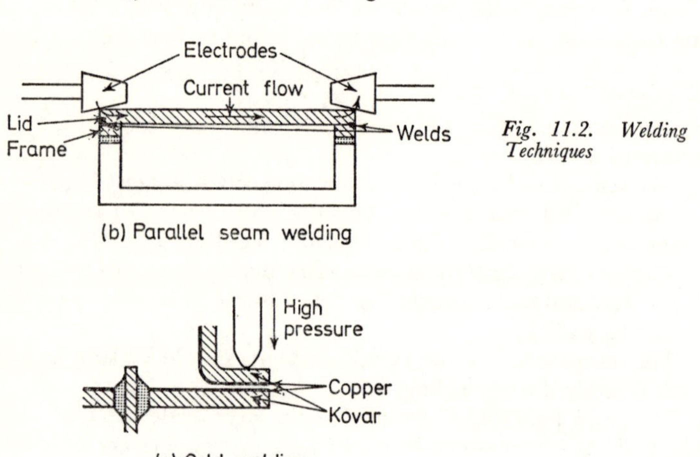

(b) Parallel seam welding

(c) Cold welding

Fig. 11.2. Welding Techniques

Parallel seam welding is more suitable for the larger metal body and closure type of packages in both circular and rectangular configurations. In this case, high frequency current pulses are passed between opposing conical electrodes which rotate along opposite surfaces of the metal lid, the latter resting on a corresponding metal frame. Current flow takes place across the lid, sufficient heat being generated at the contact areas between the electrodes and lid to melt small areas of the latter. This produces a series of overlapping molten weld areas which fuse with the metal frame to form a continuous seal.

In the case of circular packages the assembly is rotated through 180° during sealing. Square or rectangular section packages are sealed on two opposite sides by moving the assembly beneath the electrodes, followed by rotation through 90° and sealing on the remaining sides. The conical shape of the electrodes ensures that overlapping seals are produced at the corners of the package.

Owing to the localised pulsed heating employed, internal temperature rises are generally of the order of 10–50 °C only. Since the electrodes require access to one side only of the seal area, it is not necessary here to provide flanged structures, as in the case of projection resistance welding.

In cold welding, flanged components are compressed in dies in a hydraulic press to cause cold flow and weld formation without the use of external heat. To obtain the required ductility, copper-clad Kovar components are normally employed, the weld being formed between opposing copper surfaces.

While this technique affords the minimum risk of introducing contaminants during sealing, it has the disadvantage that specially designed package constructions using sealing flanges and special materials must normally be employed. Moreover the high degree of distortion which may occur during welding introduces the risk of damage to the glass feed-through seals which are employed in this type of package.

In brazing techniques, the metallic components of the package are sealed by means of a low melting point alloy which, on melting, wets both components and forms a seal on subsequent cooling.

Two alloys are commonly used, namely lead–tin–silver (92.5% Pb, 5% Sn, 2.5% Ag–melting point 280 °C), and gold–tin eutectic (20% Sn, 80% Au–melting point 280 °C). These materials are employed either as preforms or as coatings upon the metal components.

The brazing process can be carried out in a furnace, or alternatively parallel seam or hot peripheral sealing techniques may be used.

Parallel seam brazing is carried out in a chamber which, by evacuating, purging and back-filling, provides a pure, controlled atmosphere within the package. The process utilises similar equipment to that previously described for parallel seam welding (Figure 11.2 (b)). However, a relatively low current is required, sufficient only to melt the braze immediately below the electrodes. Heating of the circuit components is therefore minimised, the rise in temperature generally being of the order of 10–30 °C only. Moreover, since the lid and seal frame remain relatively cool during the process, the degree of residual strain in the package at room temperature is relatively small.

In hot peripheral sealing, which is also carried out in a controlled atmosphere chamber, a resistive heating element having the same configuration as that of the required seal area is applied to the assembly to melt the braze material (Figure 11.3). Transfer of heat to the circuit assembly is minimised by the use of a heat sink bearing on the base of the package. Compared with the parallel seam brazing method, the rise in temperature of the circuit is, however, comparatively high, temperatures of up to about 150 °C being reached. Moreover, since the entire seal freezes simultaneously, a relatively high degree of residual strain may be retained in the package at room temperature, even if it is cooled slowly after sealing.

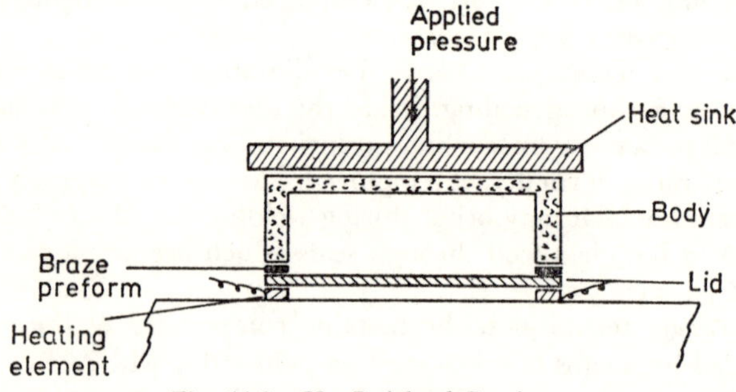

Fig. 11.3. Hot Peripheral Brazing

In furnace brazing, the package assemblies are placed in multicavity jigs under a low applied pressure and are passed through a conveyor furnace to produce the braze seal. The complete assembly must, of course, be heated to a temperature in excess of the braze melting point, with the result that some changes in circuit characteristics may thereby be caused. Although thermal shock to the circuit is avoided by a relatively low heating and cooling rate, the final package will have some residual thermal strain at room temperature owing to the simultaneous freezing of the molten braze material.

The furnace atmosphere employed is a compromise between the requirements to obtain satisfactory braze wetting characteristics and preservation of the circuit component characteristics after sealing. In contrast to other methods of sealing, the atmosphere produced within the sealed package is relatively impure, and, in combination with the high temperature attained by the circuit components during sealing, this may result in comparatively high changes in ultimate electrical characteristics. The process, however, has the advantage of being economical in high volume production.

Glass frit sealing is similar to braze sealing, but employs low temperature sealing glasses in place of metal brazing alloys. In contrast to the metal sealing methods discussed, glass frit sealing is applicable also to ceramic and glass components. The sealing glass may be of the low melting point (approximately 400 °C) with high expansion type, or the high melting point (500–600 °C) with low expansion type. The high expansion glasses will tend to cause increased residual strain in the sealed package. This yields a package which is more susceptible to shock than the low expansion glass seal types, whereas the advantage of the latter in this respect is offset by the higher sealing temperatures required.

Glass frit sealing is carried out either by furnace heating or by hot peripheral sealing. The latter technique offers the advantage that the high sealing temperatures are largely restricted to the seal areas, so minimising the rise in temperature of the circuit components. Glass frit seals, while inexpensive, have the disadvantage of comparatively low mechanical strength.

Lastly mention may be made of glasses of the recrystallisable type, which have become available for use in glass frit seals. Utilising these materials the seals can withstand considerably higher reheat temperatures; for example, the sealing temperature may be of the order of 360–380 °C whilst the reheat temperature can be as high as 450–510 °C.

REFERENCES

[1] Brauer, J. B., Kapfer, V. C. and Tamburrino, A. L., 'Can Plastic Encapsulated Microcircuits Provide Reliability with Economy?' *Microelectronics*, Vol. 1, No. 1, 5–24, (Autumn, 1970).

[2] Eleftherion, M. P., 'Assembling Beam Lead Sealed–Junction Integrated Circuit Packages.' *Solid State Technology*, 27–36, (August, 1968).

[3] Hull, J. L., 'Plastic Encapsulation of Hybrid Microelectronic Circuits', *Proceedings of Conference on Micro Electronic Circuits*, 1–8, (August, 1968).

[4] Hirsch, H., 'Resin Systems for Encapsulation of Microelectronic Packages', *Solid State Technology*, 48–55, (August, 1970).

[5] Cole, G. R., 'Hermetic Seals by Thick Film Techniques.' *Proceedings, Electronic Division of American Ceramic Society*, (1967—Fall).

[6] Bower, F. H., 'Hermetic Sealing of Integrated Circuit Packages', *Solid State Technology*, 56–61. (August, 1970).

12

Circuit Design Concepts

In previous chapters we have discussed the various components, namely substrates, passive elements, active devices, leads, and encapsulation, which make up a thick film hybrid circuit. Utilising the information gained, we shall now consider the manner in which the circuit diagram is translated into the final thick film hybrid circuit.

12.1 INITIAL EVALUATION

The major steps in a typical design sequence are outlined in Figure 12.1. The first step is, of course, to decide upon the suitability of the circuit for translation into thick film hybrid format, particularly in relation to the constraints which the thick film technique may impose on the system. For this purpose, a detailed component list is prepared from the circuit diagram. This should include the desired electrical characteristics (component values, power dissipations, temperature coefficients, tolerances, voltage ratings, etc.) for all resistors, capacitors, active devices, inductors and other components involved. The list should also indicate whether each component can be fabricated in thick film form, or must be attached in discrete form.

The choice of any component for fabrication in thick film form will, of course, depend upon a knowledge of the electrical characteristics of the various types of commercially available thick film compositions, and in particular of those which may already be in normal use in the production facility. A choice has to be made between screened low value capacitors, or the use of chip capacitors for the whole range of values. In the case of active semiconductor devices, the use of packaged or non-packaged chip

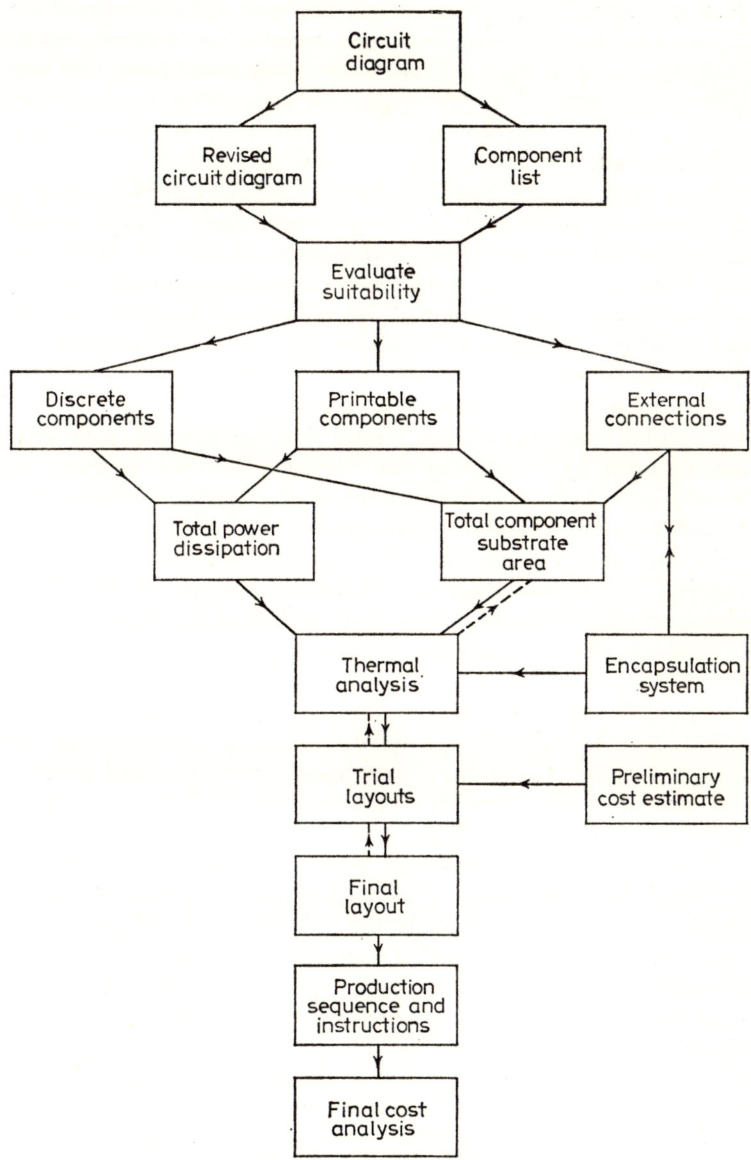

Fig. 12.1. Design Sequence

devices will be dictated by the intended application, permissible costs and, if the latter type are to be considered, the capability of the production facility for handling, bonding, and hermetically encapsulating such devices. Lastly, the components list should also indicate any items which are so large, or dissipate such power, as to warrant external connection, rather than direct inclusion in the hybrid circuit.

The circuit diagram must also be analysed and modified to make it more compatible with thick film techniques. For instance, terminal points are best grouped at appropriate edges of the circuit, matched components and resistor networks will be grouped together, and any components dissipating high powers will be separated from one another. Conductor crossovers will, as far as possible, be eliminated by suitable re-arrangement. Figure 12.2 illustrates the modification of a typical circuit in this context.

From the components list and revised circuit diagram, a decision can be made as to whether the circuit is in fact suitable for translation to hybrid format. A number of guidelines can be used as an aid here, but it should be stressed that these are general criteria only, and that the designer will in addition draw on his own experience and knowledge of the technical scope of his hybrid circuit production facility when making his final decisions.

Naturally, for economy the number of printable components in the circuit should be as high as possible, although this must, of course, be considered within the limits of substrate size and normal tolerances. As a guideline, a typical circuit may contain some 20 components/in^2 of substrate; these may comprise printed resistors and capacitors only (together with interconnections) or may include up to, say, eight discrete chips.

Fired resistor tolerances should be as wide as possible (preferably 15–20%) and rather than individual trimming of each element, this operation is best confined to the functional trimming of one or two resistors. If individual trimming is necessary, final tolerances should preferably not be set closer than 1%. The number of printed capacitors is best limited to some eight/in^2 and their production should preferably not require more than two dielectric compositions. The number of crossovers is usually limited to say, six, unless of course production capabilities and circuit complexity justify the use of a multilayer system. Whilst maximum use of screened components is advantageous, it must be remembered that the inclusion of too many components in the circuit may reduce the ultimate yield and thus increase costs. From this point of view, it is sometimes

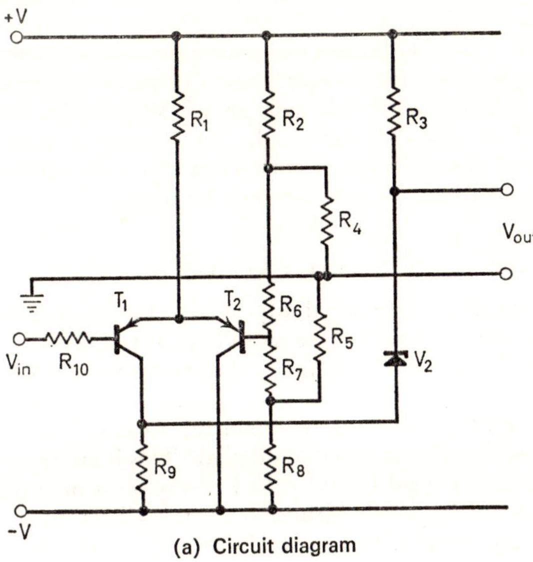

(a) Circuit diagram

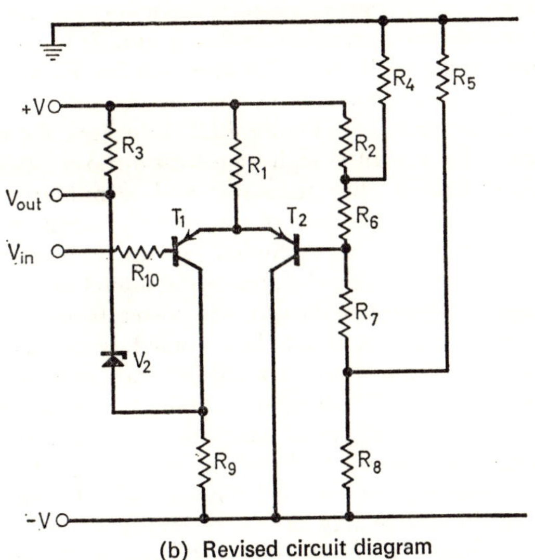

(b) Revised circuit diagram

Fig. 12.2. Preliminary Revision of Circuit layout

worthwhile subdividing the circuit into two units on separate substrates.

Depending on the thermal dissipation capabilities of the assembly after encapsulation, the average power dissipation of a hybrid circuit is about 1–2 W/in^2 of substrate area. Within the limits of the proposed substrate dimensions, the total power dissipation of the components should therefore not exceed this level, unless substrates of beryllia or special heat sinks are to be used.

12.2 TRIAL LAYOUT DESIGN

Having established the suitability of the circuit for fabrication in thick film hybrid form, the preparation of a trial layout can now be commenced. The first steps are to determine the actual area to be occupied by each component and, having regard to the required power dissipation for each component, to carry out a thermal analysis of the preliminary circuit design. This is an iterative process and the area occupied by individual components may well have to be increased subsequently with appropriate adjustment of the layout.

The layout criteria to be established, which of course must be compatible with the processing equipment and capabilities of the production facility, should relate the geometry and overall dimensions of the layout to the printed resistors, capacitors and conductors, as well as discrete devices comprised in the circuit. It is useful for the designer to derive for each type of component its 'footprint' area, that is the total area occupied by the component including its terminal conductor pads etc. Figure 12.3 illustrates, for example, typical footprints for a printed high aspect ratio resistor (see section 12.2.1), a wire bonded chip transistor and a miniature plastic encapsulated transistor. The sum of the footprint areas determines the minimum substrate area occupied by the components and terminals, excluding, however, the area of the interconnections.

The next stage is to correlate the specified power dissipation of each component with the substrate area to be occupied. In the case of printed resistors, the overall dimensions should be made so as to conform with the normally applicable dissipation figure for resistance films of 20–50 watts/in^2. The total power dissipations and areas of all the components must then be summed and correlated with the practical substrate dissipation capabilities of 1–2 watts/in^2. If this figure is exceeded, the substrate size may have to be increased or, if this is not feasible, the circuit design must be modified accordingly. In the case of circuits having a high component density or a number of high power devices, it may be necessary to employ heat sinking components such as metal fins, or to employ beryllia substrates having high thermal conductivity, in order to increase the power dissipation capability of the circuit.

It may well be advantageous in addition to undertake a more detailed thermal analysis of the circuit. Several approaches to this problem, for example, involving theoretical, analogue or experimental methods, are possible.[1–5] In the theoretical approach, thermal transfer equations are derived and solved by computer

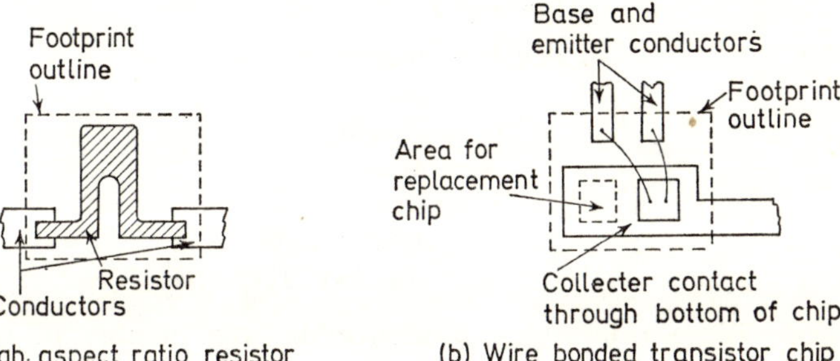

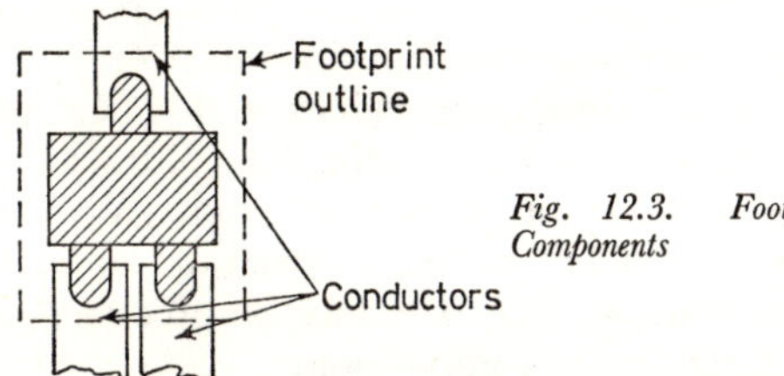

Fig. 12.3. Footprint Areas of Typical Components

analysis. Alternatively a simplified electrical circuit analogue of the thermal network can be set up employing resistors, capacitors and sheet conductor materials; voltages are then measured at points within the circuit to derive the thermal analogue.

A graphical-mathematical analysis for the determination of resistor hot-spot temperatures can also be made employing graphs and mathematical equations derived from previously established experimental data. At a later stage in the design sequence, when trial thick film layouts have been fabricated, experimental thermal anaylses can also be carried out employing an infra-red microscope, heat sensitive paints, liquid crystals, or fine-wire thermocouples for the measurement of hot-spot temperatures.

In the following sections, the design calculations for each type of component are considered in more detail.

12.2.1 RESISTORS

The basic relationship for a bulk resistor, $R = \rho l/A$, can, for rectangular printed resistors in which film thickness remains approximately constant, be reduced to

$$R = \rho_s l/w = \rho_s N$$

where R = resistance
ρ = resistivity
l = length
A = area
ρ_s = sheet resistivity
w = track width
N = number of squares

To reduce complexity in production, the number of resistor compositions employed to cover the required resistance range does not generally exceed three, since each will require separate artwork, screens and printing operations. The compositions are accordingly selected so that their sheet resistivities (specified in ohms per square) cover the widest range of resistance values within the practical limits of track aspect ratio. The latter is the ratio of track length to width (l/w), that is, number of squares of track (N), and may normally be in the region of 1:3 to 10:1. Ares istor range of about 3 $k\Omega$ to 100 $k\Omega$, for example, can therefore be produced employing a nominal 10 $k\Omega$ per square composition.

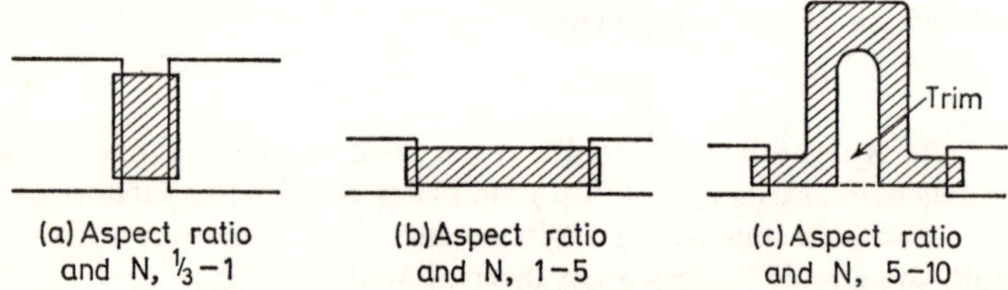

Fig. 12.4. Resistor Aspect Ratio Configurations

Figure 12.4 illustrates various aspect ratio geometries which can be used to produce printed resistance values below or above the nominal surface resistivity of the resistor composition. The 'top-hat' shaped configuration (Figure 12.4 (c)) if often used in the case of aspect ratios of between 5:1 and 10:1; if the resistor is to be trimmed, the cut is made as indicated in this case. A zig-zag configuration is not usually employed, in view of the risk of hot spots developing at the corners, and the difficulty in trimming.

If small resistors or wide variations in aspect ratio for any one resistor composition are to be used, the effect of resistor dimensions on sheet resistivity, TCR, VCR and noise should be considered. Such effects are functions in part of the interaction between the resistor and conductor films at the termination and will thus depend on the areas of overlap in these regions. The effects also depend upon the cross sectional shape of the film, that is upon the ratio of the rounded edge areas of the track to the full thickness central areas. This ratio will be greatest, and hence the effects of resistor characteristics most marked, in the case of narrow tracks[6–11].

Whereas the shape of the resistor track is determined by its resistance value, the overall size is governed by considerations of power dissipation and stability. Thus, the area of the resistor, A, required to dissipate the heat generated is given by $A = P/P_r$ where P = power dissipated (watts) and P_r = rated power (watts/in^2); P_r is generally between 20–50 watts/in^2.

Figure 12.5 illustrates a typical resistor/terminal configuration. Resistors are usually made as large as possible, consistent with the available substrate area. The minimum size is normally of the order

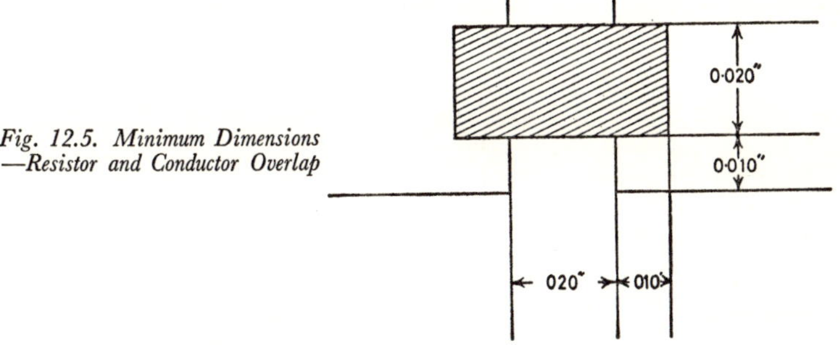

Fig. 12.5. Minimum Dimensions —Resistor and Conductor Overlap

of 0.02 in × 0.02 in, whilst the preferred resistor width is 0.03 in or greater. The resistor should preferably overlap the terminal areas by at least 0.01 in to ensure that adequate electrical contact is obtained, but to facilitate satisfactory registration in printing, an overlap of 0.015 in is preferable.

If resistors are to be trimmed by air-abrasion, a minimum clearance of 0.04 in is usually left between the track and adjacent components to allow adequate access for the abrading head. In the case of laser trimming the clearance can be less, although care must then be taken not to create hot spots by unduly close packing of resistors. If trimming is to be carried out, closed resistor loops must be avoided when designing the layout, since these would prevent the measurement of individual resistors. In such cases, a break in the loop is provided which can subsequently be closed using a jumper wire or chip conductor. To facilitate trimming, resistors are positioned to be parallel with the edges of the substrate. The reproducibility in printing will be best if all the resistors are parallel to each other.

12.2.2 CAPACITORS

In considering printed thick film capacitors, the maximum capacitance obtainable currently, using high K dielectric compositions, may be taken as in the region of 190 000 pF/in^2 (30 000 pF/cm^2) for a 0.001 in thick dielectric layer. It will accordingly be necessary, when larger capacitances are required, to utilise chip capacitors.

Screened capacitors will normally necessitate at least three additional printing stages and at least one additional firing operation. It may consequently prove less costly to utilise chip, rather than printed components, even for low capacitances, other than in the case of large capacitor arrays. A possible alternative is to redesign the circuit, eliminating capacitors either by DC coupling or by the use of zener diode coupling and decoupling elements.

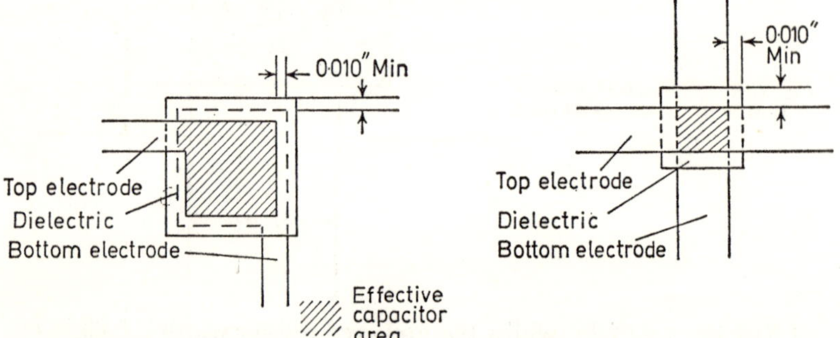

Fig. 12.6. Capacitor Configurations

The basic capacitance equation from which the area occupied by a thick film capacitor (comprising two electrodes) can be calculated is $C = KeA/t$, where C = capacitance in pF, K = dielectric constant, A = area, t = dielectric film thickness and e = constant (0.0885 where dimensions are measured in centimetres, and 0.224 where dimensions are in inches). Figure 12.6 illustrates typical capacitor design configurations in which the shaded portions represent the active capacitor area. The dielectric layer normally overlaps the lower capacitor electrode by at least 0.01 in along each edge, whilst the upper electrode falls within the lower electrode area, or crosses over the lower electrode in the insulated region. At least 0.03 in is usually allowed on one side of the capacitor for air abrasion trimming. Typical capacitor tolerances are in the range of $\pm$ 10% to $\pm$ 30%.

Whilst initial calculations of capacitor area will assume a square configuration, this can subsequently be modified to a rectangle if more suitable for the space available. The need for crossovers can sometimes be avoided by placing a capacitor at the appropriate points.

12.2.3 CONDUCTORS

Normal interconnection conductors are best made as wide and as short as possible to minimise added circuit resistance. Depending on the type of composition, surface resistivities range from 0.05—0.1 ohms per square. However, these values can be reduced to about 0.005 ohms per square by solder tinning.

A width of at least 0.02 in is preferred, although this may sometimes have to be reduced to 0.01 in or less, to correspond to semiconductor chip contacts. A spacing of 0.02 in or more is preferred between conductors, although once again in the case of semiconductor device contacts, this may be as small as 0.005 in. Conductor crossovers are avoided as far as possible, since these require additional artwork, screens, printing and firing processes (unless of course the circuit is to incorporate printed capacitors, for which these will, in any case, be necessary).

Apart from the danger of possible shorting between the crossed conductors through holes in the dielectric film, they also introduce additional capacitance in the circuit; for example, two 0.02 in wide crossed conductors represent a capacitance of the order of 2pF. If the requirement for a large number of crossovers is unavoidable, it will normally be necessary to design the circuit as a multilayer system. In this case, substrates having buried conductor layers, or multilayering techniques utilising crystallisable glass or zero-flow glass dielectrics, may be employed.

12.2.4 DISCRETE DEVICES AND TERMINAL PADS

The choice of conductor material for terminal pads here depends on whether the discrete components are to be attached by soldering, or by alternative techniques, such as thermocompression bonding, ultrasonic welding, or eutectic bonding. If flip chip or beam lead devices are to be incorporated, special fine-line compositions and printing or etching techniques will be required. The selection of conductor material is also influenced by the severity of further heating cycles which the circuit has to undergo during subsequent processing, since the adhesion of some types of conductor can be adversely affected by heating.

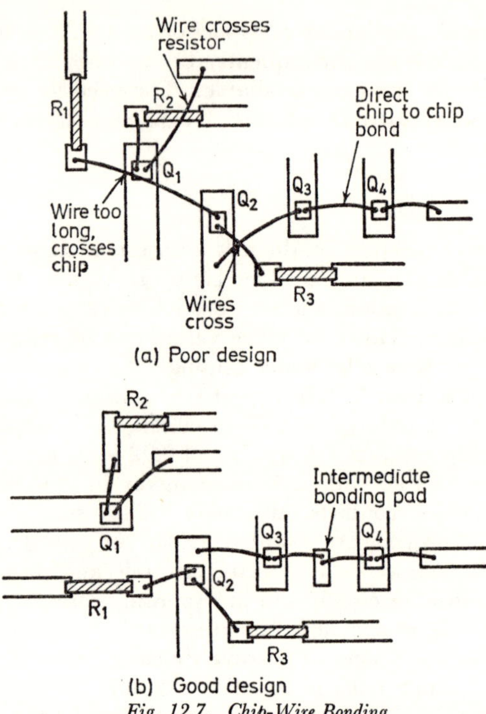

Fig. 12.7. Chip-Wire Bonding

Pads to which semiconductor chips are to be back-bonded are usually at least 0.01 in wider across each dimension than those of the chip; a 0.025 in square transistor chip, for example, will require a minimum pad size of 0.035 in square. If fine-wire interconnections are employed (normally of approx. 0.001 in diameter) the length of wire should not exceed 0.1 in.

Wire bonds between semiconductor devices should not be made directly but should utilise an intermediate bonding pad; wires should not cross over non-insulated conductors, resistors or other semiconductor chips, or over one another (see Figure 12.7). Since fine wire bonding can be a source of high failure rates, 'redundant' bonds may be provided where space permits. Typical examples here are bonds between terminal edge pads on the thick film substrate and external package leads.

In the case of conductors for the attachment of beam lead of flip chip devices, the conductor dimensions will be constrained by the small dimensions and close spacing of the device terminals; typically beam leads may be of the order of 0.003 in wide on 0.010 in centres, and flip chip bumps of 0.005–0.008 in diameter on 0.02 in centres. Typical flat pack lead widths and separations are 0.02, 0.05, or 0.1 in. Conductor widths are made as large as possible consistent with the device leads to be bonded, both to minimise the problems inherent in fine line printing, and to facilitate subsequent registration of the device leads on the conductor pattern.

In the case of capacitor chips, the contact pad areas are typically at least 0.015 in greater in each direction than the corresponding dimensions of the chip terminal areas.

Pads for the attachment of external leads are usually at least two to three times the lead diameter in width and not less than 0.05 in. in length. Chip devices are preferably mounted remote from the edge of the substrate to minimise the risk of damage during subsequent handling.

Non insulated discrete components should not be mounted directly across conductor or resistor paths; on the other hand, ceramic chip capacitors and ceramic bodied semiconductor device packages, being well insulated, may be mounted in this fashion. Care must nevertheless be taken to ensure adequate clearance between the printed component and the terminations of the mounted device, a minimum of 0.015 in normally being provided.

12.3 PREPARATION OF TRIAL LAYOUTS

Utilising the various design guidelines described in the previous sections, a trial layout can now be prepared. This can, for example, be drawn at x10 full size on a one-tenth grid using different colours to indicate various types of components. Alternatively, scale cut-outs of the footprint areas of each component can be utilised, these being placed on the circuit layout diagram and moved around to provide the best fit. In an optimum design, most of the substrate area is usefully employed, and, accordingly, the smallest standard substrate size is initially selected.

At this stage the optimised design can be checked for satisfactory electrical operation and absence of such effects as ground loops, by the construction of a simulated layout employing bread-board components, interconnected by hand wiring.

As a further stage, screens can be prepared and prototype thick film hybrid circuits fabricated and evaluated, particular attention being paid to thermal characteristics; experimental thermal analyses may be made at this point, employing the methods already indicated.

Subsequent revisions and iterations of the design sequence can now be made to produce the final layout. Sophisticated methods may at this stage utilise computer aided design and layout facilities, making use of numerical constants derived by previous experience and augmented by complex design equations, rather than relatively simple approximations.

Having produced the final layout, production instructions can now be drawn up[12]. These will comprise detailed drawings for the preparation of artwork and screens, material and component specifications, printing and firing sequence instructions, component bonding instructions, trimming and test specifications, as well as encapsulation instructions.

At this point, the substrate dimensions, numbers and types of discrete components to be included, and the sequence of printing and firing cycles required will be well defined. It will, therefore, be possible to make an estimate of the production costs of the circuit and compare it with the order of cost originally projected. If an unacceptably large difference between the two figures is apparent, it will be necessary to consider alternative materials, types of discrete components and mounting procedures and to revise the design accordingly.

REFERENCES

[1] Guntlow, V. T., Burks, D. P. and Martin, J. H., 'Thermal Design and Packaging of Hybrid Integrated Circuits', *Electronic Components*, 1045–1048, (September, 1969).

[2] Mandel, A. P. and Vahey, P., 'Thermal Design Criteria for Hybrid Microelectronics', *Microelectronics*, 36–39, (March, 1969).

[3] Ibid, 36–39, (April, 1969).

[4] Martin, J. H., Guntlow, V. T. and Burks, D. P., 'Thermal Analysis of Ceramic Based Microcircuits', *Electronic Components*, 671–680 (June, 1970).

[5] Hatzipanagos, D. and Powers, J. H., 'Thermal Characteristics of Film Resistor Modules', *Proceedings, IEEE Electronic Components Conference, Washington*, 69, (1968).

[6] Riemer, D. E., 'The Effect of Geometry on the Characteristics of Thick Film Resistors', *Proceedings, IEEE 20th Electronic Components Conference*, 102–108, (May, 1970).

[7] Kanz, J. W., 'Resistance Anomalies in Ag-PdO Thick Film Resistors', *Proceedings IEEE, 20th Electronic Components Conference*, 109–115 (May, 1970).

[8] Jefferson, C. F., 'Geometry Dependence of Thick Film Resistors', *Proceedings, IEEE 20th Electronic Components Conference*, 209–215, (May, 1970).

[9] Peckinpaugh, C. J. and Profitt, W. G., 'Termination Interface Reaction with Non-Palladium Resistors and its Effect on Apparent Sheet Resistivity', *Proceedings, IEEE 20th Electronic Components Conference*, 520–530 (May, 1970).

[10] Loughran, J. A. and Sigsbee, R. A., 'Termination Anomalies in Thick Film Resistors', *Proceedings, Hybrid Microelectronics Symposium, Dallas*, 271–280, (1969).

[11] Herbst, D. L. and Greenfield, M., 'Voltage Sensitivity versus Geometry of Thick Film Resistors', *Proceedings, Hybrid Microelectronics Symposium, Dallas*, 345–360 (1969).

[12] Fahley, W. A., 'A Standardised Hybrid Microcircuit Design Cycle', *Proceedings, IEEE 20th Electronic Components Conference*, 503–513, (May, 1970).

13

Future Trends

This chapter endeavours to project forward present-day trends in an attempt to obtain a preview of likely developments over the next few years.

One trend which is at present clearly visible and which seems almost certain to continue is the polarisation of thick film circuit applications into the high and low ends of characteristic ranges. This has been caused largely by the taking over of the middle area by silicon integrated circuit technology. As an example, there are now major developments of thick films in single and multilayer conductor patterns, where they act as mother boards for integrated circuits, and also in the fabrication of short runs of highly complex and specialised circuits. Similarly, the thick film hybrid approach has considerable advantages for high and low power requirements, and again for very high performance or performance insensitive uses. In each case the middle part of the range seems likely to continue to be dominated by silicon integrated devices.

A factor which may further strengthen these trends is the increasing availability of both active and passive discrete devices in forms convenient for attachment to thick film hybrid circuits and multilayers. It appears that thick films will maintain this advantage in relation to integrated circuits for quite some time, since there are no indications that it will be possible either to attach discrete components to the latter or fabricate integrally, for example, capacitors, with much improved characteristics beyond those already available.

An additional capability which could extend the range of application of thick film circuits very radically would be the ability to produce thick film active devices by methods compatible with existing screen printing technology. Up to the present time, devices produced with, for example, cadmium sulphide/selenide, have proved to be rather unattractive from the performance point of view; however, we may well see some marked advances in this area over the next few years. Similarly one can expect to see various other components produced in thick film format and indeed, development work on diodes, amorphous semiconductor devices, thermistors, etc. is already under way.

Further ahead, there is the possible development of a wide range of magnetic, optic, ferroelectric and similar devices in a form suitable for addition to thick film circuits, but, at the present time, it is difficult to envisage how these could be produced in an integrated circuit matrix. A possible exception here could be, for example, a hybrid silicon/gallium arsenide/crystalline ceramic technology, but this does not appear likely for some time.

On the materials side, there will become available low cost conductor and resistor pastes which are less critical in respect of processing conditions, and probably also, with appreciably lower firing temperatures. These factors would, in turn, make possible the use of a wider range of substrate materials. There will also be considerable improvements in dielectric and crossover pastes. These pastes may be raised to a similar standard of performance and reproducibility now shown by resistor pastes; in particular, reliable pastes with higher permittivities should become increasingly available.

It seems likely that use will be made of a much wider range of substrate materials, due to the availability of lower temperature processing pastes, and also to the commercial pressure for lower materials costs. It should, for example, prove possible before long to produce flexible thick film circuits, with considerable space saving for difficult geometries.

One might also expect to see a progressive displacement of the printed circuit mother board by thick films on ceramic or insulated metal substrates, where the capability for higher power dissipation is required and where the attachment of unpackaged chips at high density is desirable. A further development in this area which may come about for the latter requirement is the extension of the so-called selective etching techniques to thick film devices to produce particularly intricate and precise patterns. This, for example, will be increasingly required for the mounting of LSI and SSI semiconductor chips.

In the field of discrete components, increasing use of thick film materials processes as a means of forming resistors and capacitors may be expected. Probably they will also be used as the basis for conductor patterns, electrodes and contact areas in devices such as switches for high temperature applications.

It would seem likely that the major expansion in the volume of thick film circuit production will come from an increase in the number of small-to-medium sized 'in-house' facilities, rather than a few large general suppliers to the industry, since in the former case the special advantages of economical short runs and ready adjustment of circuit variables are most usefully realised.

Also in favour of the 'in-house' approach, is the fact mentioned previously that thick film circuits are probably most attractive in the fabrication of small numbers of relatively complex circuits, possibly including multilayer interconnection patterns which may carry a range of specialised components or functions at relatively high density and cost. A major part of production costs in these cases will be in the setting up and operation of the necessarily complex test facilities, since it will be desirable to test these circuits as systems rather than as individual components. Thus, the greatest advantages in this area would appear to accrue to thick film circuit producers who are also users; for example, manufacturers of instrumentation and control systems who already have complex test gear available for performance testing of these very same functions.

On the other hand there are other interested parties who may seek to enter thick film manufacturing on a large scale. The semiconductor manufacturing houses who, as their devices grow in complexity and packages become larger and more multi-leaded, must increasingly come to regard the hybrid microcircuit carrier substrate which connects their product to the outside world as a vital part of their product development. As an example of existing involvements in this area, there is the present trend of multi-chip MSI (pending development of true monochip MSI) leading to a multi-MSI package, in the course of the development of true SSI (super scale integration). Having arrived at this stage where semiconductor manufacturers are taking over the yield-crucial stage of mounting the complex and expensive chips onto thick film connection patterns, it would seem a logical step for the manufacturer to incorporate upon the mother ceramic board those other discrete components which may be necessary to complete the functional system but which yet are not capable of being integrated into a conventional monolithic silicon semiconductor system.

There is, of course, always the possibility that a completely new technology will arise from within the present thick film trends, but this appears unlikely in the near future. It seems certain that for some time ahead there will be a considerable expansion in the use of thick film hybrid microcircuits in the areas discussed. This should be largely in applications which capitalise on the capability of the technology to provide a substrate, interconnect and packaging system for specialised complex circuits. These would contain a large variety of both active and passive devices with a combination of performance characteristics not feasible within the compass of monolithic semiconductor technology alone in the foreseeable future. Thus, hybrid thick film technology is expected to become increasingly utilised as a basis for circuit fabrication across the whole spectrum of electronic/electrical applications.

Index